The FD3/67 LS turbojet is the latest development of Kurt Schrecklings motor, and is assembled from a kit of parts with no special tools being necessary. (See chapter 10)

Gas Turbines for Model Aircraft

Gas Turbines for Model Aircraft

Kurt Schreckling
Dipl. Ing.

A Traplet Publication

Translated from the original German by Keith Thomas

Traplet Publications Limited
Traplet House
Severn Drive
Upton upon Severn
Worcestershire
WR8 0JL
United Kingdom

ISBN 0 9510589 1 6

Printed and bound by Stephens & George Limited
Merthyr Industrial Estate, Dowlais, Merthyr Tydfil, Mid Glamorgan CF48 2TD

Cover: Mike Koskela's FD3/67LS powered BAe Hawk, and inset, Arthur Griffiths home built FD3/64 (Mike Cherry photos)

About the Author

(Photo Dr. Gerhard Rubin)

Kurt Schreckling, Dipl.-Ing., born 1939

Kurt Schreckling's education began with basic technical studies, after which he completed an engineering course with the emphasis on applied physics. Since then he has worked for a large chemical company in process control and system technology.

Before he reached his fifth birthday Herr Schreckling had his first practical experience of model flying when he converted a tangled kite into a model aircraft. Some years later he started building model aircraft and radio control systems. His particular love was power systems, which at that time were not yet highly developed. In consequence he invested considerable time in the development of electric flight: variable-pitch propellers and computer-optimised electric flight power systems. Then came his first successful record attempt, using an electric helicopter he developed himself, followed by the design of the electric power system for Wolfgang Kueppers' first record-breaking speed model. For five years he has devoted most of his spare time to the development of the model turbo-jet engine. He also finds time to write about his success in the field.

As a result, when the decision was made to develop a professionally produced kit for a model turbo-jet, Herr Schreckling was the obvious choice of collaborator.

Herr Schreckling does not claim to be a particularly good model pilot. However he has ingenious ideas, and he more or less single-handedly develops them to practical form, then installs his engines in models and goes off and flies them, he must be classed as one of the most versatile and experienced modellers of our time.

Foreword to the second edition

To date many successful variants of the FD 3/64 have been constructed, and this led me to consider adding an appendix to this new edition of the book, covering a range of specialist matters relating to the model turbo-jet. On the other hand, if I had taken a thorough approach to this appendix, the result would certainly have been outside the scope of the book, and might even have confused people. Many questions have been addressed to me along the lines of "why did you design the FD 3/64 like this, and not like that?", and I will only ever be able to supply partial answers. When faced with a pressing problem, such as the oil supply to the bearings, I tried to come up with a simple, practical solution, rather than to carry out exhaustive tests on other, possibly better, systems.

Reiner Eckstein won the coveted "Best of Show" award at the first Ohain-Whittle trophy competition, with this own-design 'Turbo-Trainer', powered by a homebuilt FD3/64.

Many modellers have been successful with turbo-jet models, and their activities came to a climax in June 1994 with the Ohain/Whittle Trophy competition held in Nordheim. In spite of competition from semi-professional modellers, the "Best of Show" trophy was won by a Turbo Trainer powered by an FD 3/64, built and flown by Reiner Eckstein.

Since the appearance of the first edition many real developments have taken place, and extensive flying experience has been gained with semi-scale models and FD 3 engines, not to mention the development of a kit for the FD 3/67 LS turbo-jet, now a refined and fully developed design.

Naturally I will always be available for comments and suggestions relating to problems with engines built from my drawings. To the many modellers who have listened patiently to my telephone sermons in the past, I would like to express my grateful thanks.

Kurt Schreckling

Contents

Chapter 1

Introduction

This turbine wheel, powered by thermal energy, was the very first stage in the development of the model turbo-jet.

A turbo-jet engine for model aircraft? What's a turbo-jet, anyway? "What sort of engine have you got in there, then?" These are the questions I have heard so often, even from experienced power modellers. Sometimes a little more knowledge was betrayed by the question: "How many turbine wheels has it got?" Occasionally – very occasionally – I would be asked a question on the compression ratio. Then I knew that I was in the presence of a real expert! But all my interrogators had one thing in common: they all wanted to know exactly how this jet turbine device worked.

A turbo-jet, also known as a jet turbine engine, exploits a gas turbine to produce thrust. It is called a gas turbine because the working medium – air – is in a gaseous form. Please note that this has nothing to do with the possibility of using fuel in a gaseous state. In its simplest constructional form this type of heat engine makes a high-performance power source for an aircraft. The gas turbine becomes a jet turbine, or turbo-jet, when the usable energy in the exhaust gas from the turbine is concentrated using a nozzle, or jet. However, this is not essential in principle.

The first aircraft propelled by a jet turbine was the He 178, which flew for the first time on 22.8.1939. It was built in the Heinkel factory and the pilot was Erich Warsitz. This revolutionary engine was the creation of Dr. Papst von Ohain. On its very first flight this aircraft – the first ever to be powered by a turbo-jet engine – reached a speed of 600 km/hr, faster than any other series-produced airscrew-driven aircraft of the time. In the truest sense of the phrase the engine was simple and ingenious: a single compressor wheel sucked the air in and compressed it in the combustion chamber. Heat energy was added by burning fuel, and the hot

The first experimental design, aimed at proving that a turbo-jet constructed using simple means could be made to work, was completed in April 1989 and ran under its own power on petrol.

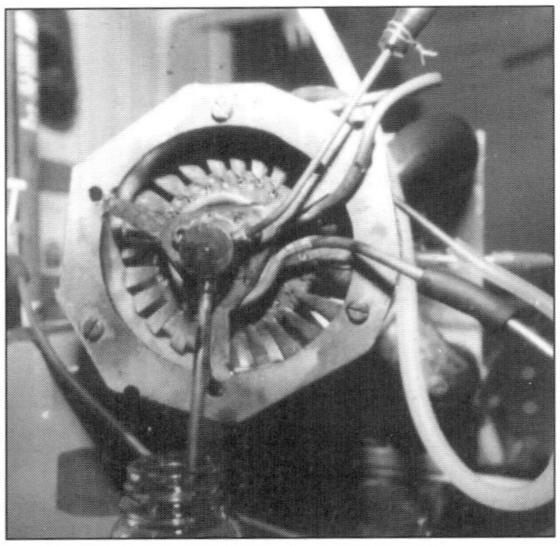

A view of the turbine end of the experimental engine. The shaft bearing is external, and the bearing was cooled with oil and a coil of tubing. This idea did not prove satisfactory (Photo: Reiner Binczyk).

View of the diffuser system with the turbine wheel removed (Photo: Reiner Binczyk).

exhaust was ducted via the turbine to the open air to produce thrust. The turbine wheel extracted just enough energy from the hot exhaust gas to drive the compressor. All modern turbo-jets, turbo-props, turbo-fans and the closely related engines used in many helicopters are based essentially on gas turbines, which by now have been developed to a very high level of sophistication. Dr. Papst von Ohain would hardly have recognised his simple basic principle in the engines of today.

The modeller's desire for a real turbo-jet for his model aircraft is perfectly natural, and there is evidence enough in the many semi-scale models of jet aircraft which are popular today. However, for many years it seemed extremely unlikely that the dream could ever become a reality. One thing is certain: making a reduced-scale replica of a modern turbo-jet engine would never produce a working model jet turbine. For one thing the full-size engines are extremely complex in design and construction, but in any case the laws of physics are weighted against us. Nobody would think of trying to make a scale working model of a twin radial piston engine as a viable model power plant, and full-size turbo-jet engines are even less suitable for shrinking to model dimensions. If you take the design of a small piston engine and make it smaller and smaller, or distribute its swept capacity over many small cylinders, its power output steadily diminishes, and the complexity of its construction steadily increases, with the result that some engines of this type cannot be persuaded to run at all. This problem is widely acknowledged, and although these engines may well be beautiful examples of miniature engineering, they are doomed to a life in a showcase. A number of model jet turbines have been built in the form of miniature full-size engines. Their sole disadvantage is that they don't work. It is also well-known that the flying characteristics of true-scale models of high-performance aircraft deteriorate as the scale is

The compressor end with the cover open and the compressor wheel removed. The housing has four diffuser outlets and no diffuser vanes. The system is quite efficient, but too bulky for its purpose as a model aircraft power plant (Photo: Reiner Binczyk).

Producing a turbo-jet capable of flight was not that easy, and the "FD 1" (left) was not the answer. It was capable of running autonomously, but the operating temperature was much too high. A re-designed compressor with a larger diameter housing resulted in the "FD 2".

reduced. Nevertheless it is perfectly possible to build very small gliders which fly well, and small piston engines which are very powerful. There are also a few – a very few – tiny jet turbines which do work. All this is possible using relatively simple techniques and technology, provided that you observe the laws of physics and exploit them correctly. Regrettably there is one idea which is just not true: that reducing a machine to scale is bound to produce a working replica.

Until now there has been a complete dearth of literature concerning the technology and theory of model jet turbines. This fact, together with my successful exploits in the field and the many requests I have received for information, persuaded me to write this book. Its purpose is to provide you, the interested reader, with all the information you need to build your own model turbo-jet engine, both in terms of basic physics and practical technology.

In preparing this work I have taken into account the fact that the majority of modellers are neither engineers nor physicists. However, I do believe that it is essential to impart a certain level of understanding of physics and technology as they apply to this revolutionary type of model power plant before I present the building instructions and associated technical drawings. Chapter 2 therefore attempts to instil understanding without onerous formulae and mathematics. Chapter 4 offers a theoretical grounding relating to the practical requirements of the model turbo-jet, aimed at the reader with some degree of technical and/or scientif-

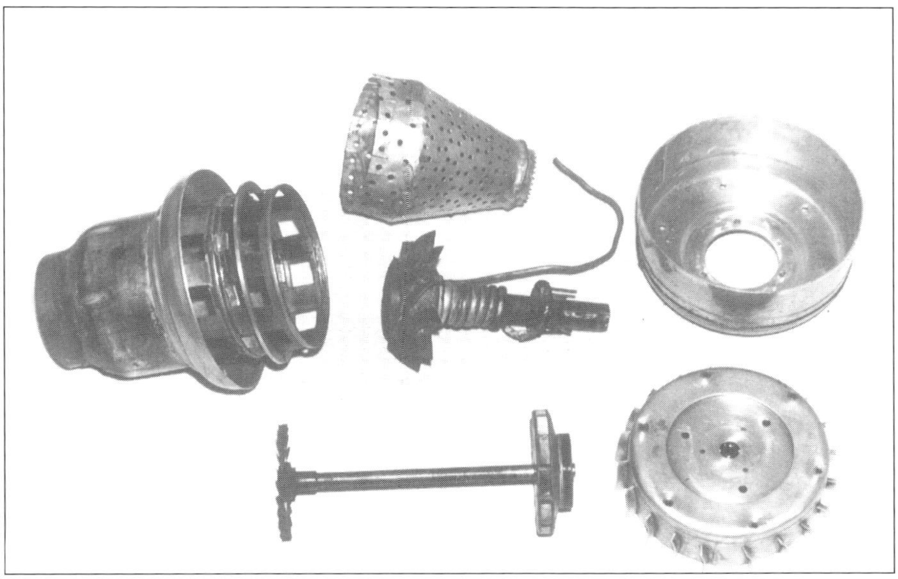

Above: The original version of the "FD 2" dismantled. The combustion chamber and shaft shown here had to be re-designed before the engine was capable of flight.

Left: The turbine end of the "FD 2". With the turbine wheel spinning we can see the centre of the turbine disc and the diffuser system vanes (Photo: Dr. Gerhard Rubin).

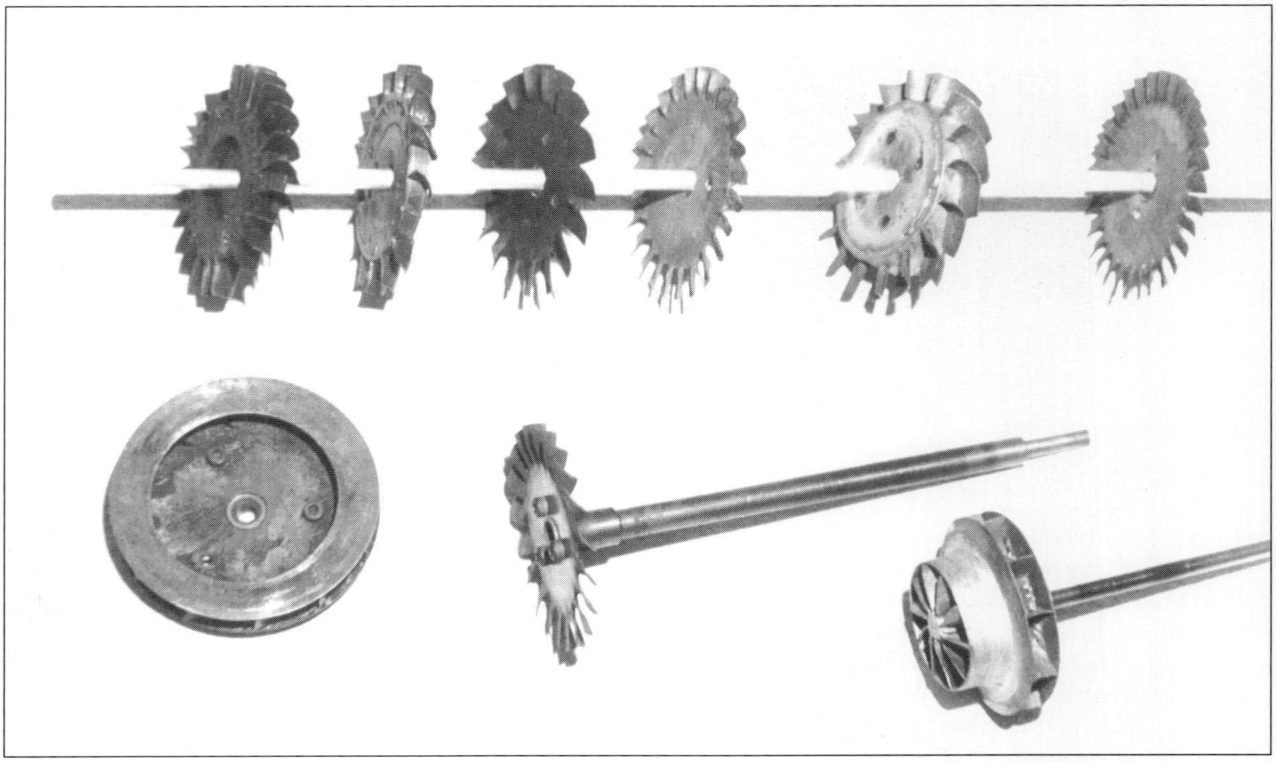

Experiments with various turbine wheels led to the surprising conclusion that the simple form, similar to that shown in the first photo, is entirely satisfactory for the requirements of the model jet turbine.

ic knowledge, although I have not attempted to attain scientific perfection. The jet turbine described in the building instructions is the outcome of my own development work, and is a practical engine, proven in the air, and built using nothing more in terms of tools than is to be found in any well-equipped amateur workshop. I have tried as hard as possible to simplify and simplify again, but even so the engine is a highly technical machine and must be made to a good level of precision. Please don't underestimate this! Your workshop must be equipped with at least the following items:

1. Lathe, at least 54 mm centre height, 300 mm between centres
2. Gas welding (MIG) apparatus
3. Hard soldering equipment with oxygen support
4. Accurate pillar drill for boring holes in the range 0.5 mm diameter to 10 mm diameter
5. All the usual metal-working tools, such as drills, files, saw, hammer, pliers and measuring tools
6. Measuring equipment for operating the jet turbine, such as tachometer, thermometer, manometer, and a thrust measuring device.

In contrast to the tools required, the materials are much less expensive. When selecting materials I deliberately kept general availability in mind at all times. The parts list which supplements the building instructions, and the list of sources of supply should make this clear. However, of all the essential ingredients of a successful home-constructed turbo-jet, the most important one is the manual skills of the builder. If you are a modeller who above all loves to experiment, I urge you to read this book right through before you launch your own independent line of development. On the other hand, if you are content to follow the building instructions to the letter, it is not absolutely essential to understand the theory in full. But even if you consider yourself a seasoned practitioner, please read through the theory!

The first series of pictures documents the stages of development from the first experimental engine to the successful "FD 3/64" turbo-jet, which has been extensively flight-tested. I find it very rewarding that my work has encouraged other modellers to start similar projects. One of the more successful ones is that of Reiner Binczyk. Our highpoint to date was the simultaneous flight of Reiner Binczyk's turbo-jet powered model and my "Rutonius" at a demonstration event at Holstebro, Denmark, on 24.8.1991.

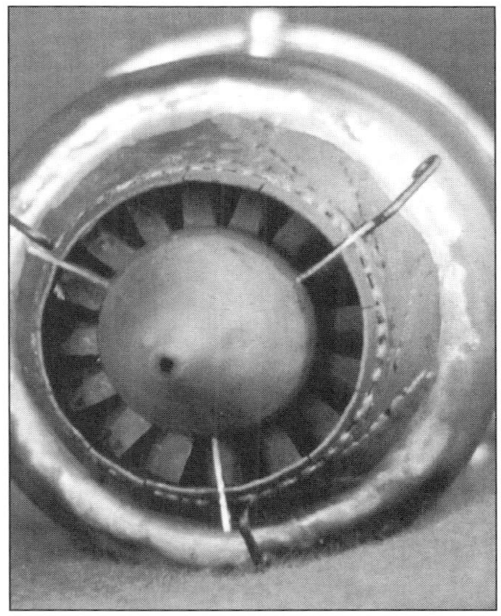

The "FD 3/64" in its latest state of development. The only difference in the compressor wheel compared with the first experimental version is its carbon fibre reinforcement. The turbine wheel is also thicker. Static thrust can be optimised by adjusting the annular jet.

Diagram of a turbo-jet engine

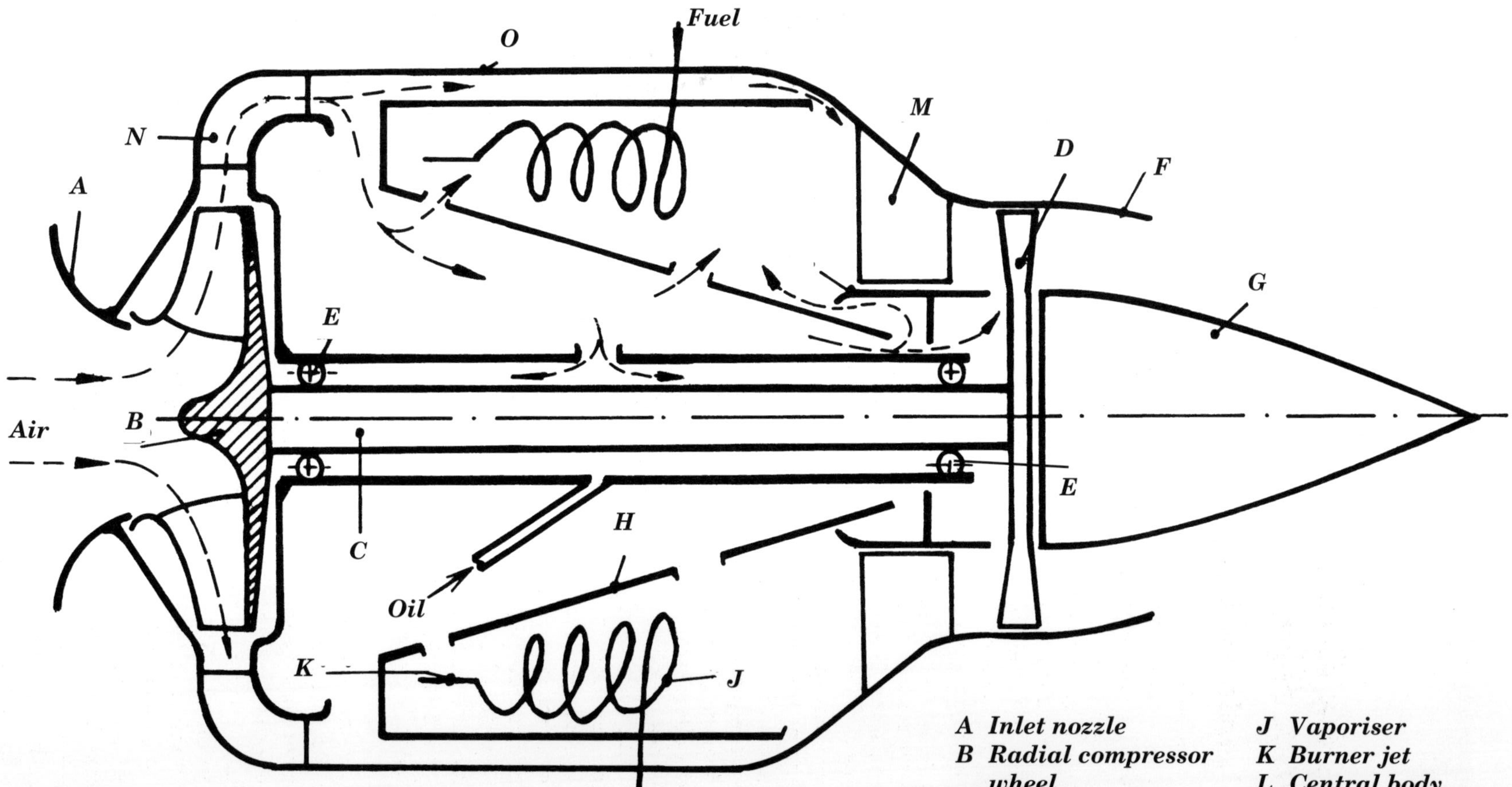

A Inlet nozzle
B Radial compressor wheel
C Shaft
D Turbine wheel
E Ballrace
F Exhaust jet
G Flow stabiliser
H Combustion chamber
J Vaporiser
K Burner jet
L Central body
M Turbine diffuser blade
N Compressor diffuser blade
O Housing

Chapter 2

The Basic Physical and Technological Principles Behind the Model Turbo-jet Engine

2.1 How a simple turbo-jet works

The heart of a turbo-jet – the gas turbine – can be classified as a normally aspirated heat engine, together with the reciprocating piston engine, the pulse jet and the ram jet. These engines convert part of the energy produced by fuel combustion into usable energy. The turbo-jet's sole source of usable energy is the kinetic energy of the exhaust stream which is emitted at high speed. The magnitude of the thrust is found by multiplying the velocity of the exhaust stream by the mass of the gas emitted. Energy conversion in this type of engine is only possible if the working medium, in this case air, is first increased in pressure relative to the atmosphere. An accurate physical explanation of this process involves an excursion into the theory of thermo-dynamics, which is well beyond the scope of this book. When reading the next section, please refer to the diagram of a turbo-jet, which is intended to illustrate how the engine functions.

One obvious illustration of the fact that an engine of this type cannot work without compression is that of a model piston engine with a blown head gasket or a loose glowplug. In a piston engine the piston works alternately as an energy consumer (compressor) and as an energy contributor, i.e. during the power stroke. However, the engine can only run autonomously if the energy produced during the power stroke is higher than the energy consumed during the compression stroke plus the additional energy expended at the shaft, plus friction losses. In comparison, the gas turbine can only run autonomously if the turbine's shaft power is equal to or greater than the energy absorbed by the compressor in the same timespace. In terms of physics the work performed per unit of time is the engine's power. In a machine which constantly executes work, we are justified in equating its work output with power.

When considering a piston engine we generally know the shaft power. No thrust is produced until we fit a propeller to the engine. Therefore it is not possible to produce a simple and direct comparison between the thrust of a turbo-jet and the shaft power of a piston engine. This matter requires further explanation, and is discussed in detail in Chapter 3.

Let us start by attempting to understand the turbo-jet's method of working a little more fully. In a turbo-jet compression and production of shaft power take place constantly. Since energy-consuming compression and work-producing decompression of the working medium – air – cannot occur at one and the same location simultaneously, the turbo-jet requires two separate stages, namely the compressor stage and the turbine stage. Each of these stages consists of a fixed system of diffuser blades plus a revolving rotor – the compressor and turbine wheels – which are coupled to the shaft. The sub-assembly consisting of shaft, compressor wheel and turbine wheel is termed the rotor. Heat energy is added in the combustion chamber, through which the whole of the airflow streams. The combustion chamber is located between the compressor and the turbine stage, and in itself is not a particularly complicated component. The gas turbine is completely indifferent to the type of fuel it is fed. However, achieving intensive combustion in the smallest possible space, as required for a practically useful model turbo-jet engine, can consume considerable effort at the experimental stage.

All gas turbines have one dangerous characteristic in common, and this must be taken into account at all times: they are insatiable fuel consumers. The more they get, the higher the thrust, temperature and rotational speed. At the same time the efficiency of the energy conversion process also increases, and with it the rate of speed increase. If the fuel supply is not restricted, the turbine's speed races away until one or other rotating part can no longer withstand the centrifugal load. This always applies – even when the finest materials are used. The inevitable result is then turbo-scrap. Yet even this problem is soluble, and we will discuss it in detail. The basic requirement is to prevent the gas turbine from racing, or "running away", as we call it.

The difference between a simple gas turbine and a jet turbine is minor, and appears at the tail end. Every gas turbine which produces a directed exhaust flow is already a turbo-jet. In a turbo-jet "proper", a jet or nozzle is fitted behind the turbine stage, to amplify and optimise the performance and thrust of the exhaust flow. This poses no particular problems in the engine's application as a model aircraft power plant.

At this point a few words are in order on starting a gas turbine. Like a piston engine it is unable to start independently from 0 rpm and run up to working speed under its own power. To start running it requires the supplementary energy of a starter. Please note that the engine is capable of "running away" at the initial start-up stage, if, for example, it is flooded with fuel. Before you actually attempt to run your engine it is therefore absolutely essential to study the operating instructions.

Supplementary energy is also required for ignition. In contrast to a piston engine, combustion in a gas turbine is a continuous process, so the mixture only needs to be ignited once. This is the least of all gas turbine problems.

2.2. The right way to construct a gas turbine rotor using amateur means

In technical terms the rotor is undoubtedly the most complex part of any gas turbine. If you can build the rotor, you can undoubtedly cope with the remaining technical problems. An obvious ploy would be to settle for a ready-made rotor from a car turbocharger, and design the rest of the engine around it. These rotors consist of a semi-open compressor wheel with radial tips and a turbine wheel of similar form. This starting point for developing a model jet turbine would undoubtedly work, but unfortunately goes beyond the bounds of the possible in an amateur workshop with the equipment listed in the Introduction. The primary problem is the construction of an accurate housing, as the configuration of housing and rotor with a radial compressor and turbine wheel is extremely sensitive to tolerances in the axial direction.

The problems involved in making a housing complete with bearings for a ready-made rotor are very likely to be more severe than those encountered in the procedure described in the following text.

The first step is to forget about all the model turbo-jets you may have seen or read about, because none of them were built using amateur tools and methods. Success has been achieved by applying the theoretical calculations and considerations outlined below, followed by painstaking practical experimentation:

1. The laws of physics apply to small gas turbines in basically the same way as to full-size engines of the same type. The only problem in this respect is the difficulty of calculating and assessing the inevitable losses – what we might term internal efficiency. However, it is possible to calculate the maximum permissible loss, i.e. the level of internal efficiency required for the gas turbine to function.

2. If we consider comparable mechanical assemblies, such as a model airscrew and a man-carrying aircraft propeller, we find the following: the maximum efficiency of the full-size propeller lies in the range 85 to 89%. My own experiments with electric flight models showed a best efficiency for a model propeller of around 75%. From this numerical comparison we can see that the crucial efficiency of the turbine engine does not show such a dramatic fall-off, in spite of the considerable reduction in scale. If we compare the airflow conditions through a propeller and through a compressor wheel, we find distinct similarities. In both cases the airflow is first accelerated and then slowed down again. With a compressor wheel fitted with radial blades (as an example) the air is sucked in and accelerated by the rotational motion of the impeller to a peripheral speed of more than 200 m/s as it flows through the wheel. Of course, power is required to compress the air in this way. Part of the pressure increase is accounted for by centrifugal force. The other part is due to the compressor's diffuser system slowing down the flow. Unfortunately we have to accept losses of around 20% in this process. Additional losses accrue due to friction and gap inefficiency as the air flows through the impeller. The laws of physics prevent us building a radial compressor stage without this braking effect.

Nevertheless, there is a particular type of radial compressor wheel in which this high-loss effect – namely the slowing of the airflow – is much less serious, and in which the gap losses can virtually be eliminated. This is the radial impeller with retro-curved diffuser blades and cover plate. This type of wheel is used in industrial air duct systems and gas supply installations, and is made in a very extensive range of sizes. These impellers achieve efficiencies of more than 80%. One familiar application in a small form is the suction wheel of a vacuum cleaner. But please don't start ripping the vacuum cleaner to bits in the hope of making a turbine compressor from it! Some readers interpreted one of my articles as if this were the case, although all I mentioned was the coincidental similarity between the first successful jet turbine, the FD 2, and a vacuum cleaner motor. Let me take this opportunity to state, once and for all, that I have never used any part of a vacuum cleaner in a jet turbine!

If I could reduce the size of the impellers used in industrial installations to the dimensions of a rotor required for a turbo-jet, and achieve a loss in efficiency comparable to that obtained with propellers, then the battle would be over. Could it be done? I was able to answer this question by experiment: I built a model compressor stage powered by a high-performance electric motor. Calibrated nozzles were set up downstream of the compressor stage. Starting from the known efficiency of the motor and its power consumption, and pressure measurements taken at the outlet nozzles, it was possible to calculate to a fair degree of accuracy both the impeller's efficiency and its characteristic curve. The results were encouraging. The maximum efficiency of these small compressor wheels is around 75%, i.e. only 25% of the shaft power is lost. The method of calculating these figures has been described in great detail in the specialist press, e.g. by Bohl (1), and these methods can be applied very well to small compressor wheels. As is the case with full-size compressors, the characteristic curve of the compressor wheel is non-critical. This is helpful to us, as it means that the engine's operating characteristics do not deteriorate significantly when load changes, which occur when the engine is running, altering the overall flow through the gas turbine. The net result is that we can expect the engine to behave in a stable, reliable manner when running. The matching of the diffuser system to the compressor wheel is also extremely flexible if retro-curved diffuser blades are used.

In the interests of completeness I ought to mention that the electric motor used in these experiments could only provide rotational speeds of up to about 20,000 rpm, which approximates to the idle speed of the turbo-jet. However, according to the laws of fluid dynamics we can expect that the flow losses will diminish in proportion to usable work as rotational speed and air throughput rise. Similar effects can be observed in the flight of models at high and low airspeeds, and can be laid at the door of Reynolds numbers, which rise at higher speeds. In general terms this means that friction accounts for a falling proportion of total air resistance.

The compressor wheel with cover plate also offers a crucial constructional advantage for our application: the permissible tolerances in axial clearance in the housing are much greater than those for the semi-open compressor wheel used in a turbo-charger, where tight tolerances are essential. If this were not the case, gap losses would have a very serious effect on efficiency, and would make it impossible to make a working model turbo-jet.

The sole drawback to the retro-curved diffuser blades is the necessarily greater diameter of the wheel compared with radial tipped blades. This means that the retro-curved wheel has to cope with higher rotational loads for a given quantity of compression work and a given throughflow and rotational speed. My development work has shown that sufficient strength to cope with the rotational loads encountered in a model turbo-jet can be achieved using plywood as the basic material, reinforced with carbon fibre. The building instructions contain full details of how to construct these parts. This technique results in a very lightweight wheel, which in turn eases considerably the design problems relating to the shaft and its bearings. It also makes balancing easier – and good balance is indispensable to a smooth-running gas turbine.

One question remains to be answered: why not use an axial compressor? And the short answer – one which will satisfy the real model engineer – is this: just try it! I don't want to get tangled in masses of formulae at this point, but I hope the following gives you the general idea: in deciding on the small compressor wheel with retro-curved diffuser blades I applied certain rules of physics and maths. If we apply the same rules to the axial compressor we find that it requires at least four stages to give comparable performance. This means constructing four compressor wheels and four diffuser systems. Full-size compressors of this type are more efficient than a radial compressor, but we cannot expect the same improvement because of the low Reynolds numbers at our diffuser blades. Professionally produced miniature gas turbines broadly comparable with our turbo-jet do exist, but none of them has an axial compressor stage. My own measurements with a small axial compressor showed that it was markedly less efficient than a radial compressor.

Now let us consider the problems at the hot end of the rotor: the turbine end. And I do mean hot. The choice of a turbine wheel with axial throughflow side-steps a number of problems relating to housing design: the only basic requirement is that the wheel must be accurately centred in the housing, and there must be clearance at the periphery. Axial play at this point is not a problem. The distance between the diffuser system and the turbine wheel is similarly insignificant in terms of the efficiency of the turbine stage. The energy in the compressed air is increased by the heat of combustion, and the purpose of the turbine wheel is to absorb part of that energy and transmit it to the compressor wheel. The remainder must be allowed to pass through unhindered, to provide the energy for the exhaust gas stream. In fact every air-operated gas turbine does that almost automatically. For our modelling purposes we hardly need to interfere at all.

The two main problems relating to the turbine wheel are the magnitude of the centrifugal load and the high operating temperature. It is obvious that power rises very considerably as rotational speed increases, and the rate of increase is even steeper than the rise in rotational speed. But at some stage the stress on the turbine wheel becomes so severe that it simply can no longer bear it – in the true sense of the phrase. It is precisely for this reason that small professionally produced gas turbine wheels are made of special high temperature alloys using precision casting methods. Neither these materials nor the manufacturing process are within the scope of the amateur workshop, and spark erosion techniques and machining using CNC equipment can be ruled out for the same reason.

However, we should not give up hope just because we cannot use the techniques described above. Let's not lose sight of the fact that we are aiming at a working model jet turbine. To help us on our way, we will anticipate one conclusion which is reached in Chapter 4.1: it is by no means essential that the operating temperature at the turbine blades should be as high as those encountered in commercially produced turbo-jets. In our case good results can be achieved with a gas temperature at the blades of around 600° C. Careful ducting of cooling air can reduce the temperature of the turbine disc and the highly stressed transitional area at the blade roots to a much lower level. Naturally, this is still far too hot for wood, aluminium and similar materials. Even normal and low-alloy steels lose so much strength at temperatures above 400° C that they are no longer able to withstand the centrifugal load at the necessarily high rotational speeds of a turbine. When dealing with the centrifugal load it is more useful for us to consider the peripheral speed of the rotating body. In this case geometrically similar bodies of different size will have the same centrifugal load if their peripheral speed is the same.

A serviceable material for the turbine wheel is nickel-chrome steel, also known under the names V 2A, V 4A and Remanit. An even better material is nickel-chrome steel alloyed with molybdenum. These materials are used in a variety of thicknesses in fitters' workshops – especially where they have anything to do with furnaces and boilers. You should find little difficulty in obtaining a few pieces of scrap material, and scrap metal merchants are a further possibility. Lay your hands on this material, and you have solved the main procurement problem for your model turbo-jet.

Now let us turn to the shaping. The materials mentioned above can be worked with standard HSS cutting tools, i.e. they can be drilled, sawn, filed, turned and ground, and can equally well be formed by bending. They can also be gas-welded and hard-soldered.

Before we get immersed in a complex project it is important to know whether we have any prospect of success. Please don't be worried by this description of my experiments! You don't need to repeat any of this work if all you want to do is build a turbo-jet. At this stage the crucial question for me was this: is it possible to produce a serviceable turbine wheel without having to build a complete gas turbine? With this in mind I carried out experiments with the aim of finding out the extent to which a turbine

wheel could be simplified in terms of manufacturing techniques, whilst still maintaining an adequate level of efficiency for a turbo-jet. Now that my experiments with the miniature compressor wheel were complete, and the calculations made and checked, the final uncertainty was the efficiency of the turbine stage. The only way of obtaining reliable figures was to use experimental methods.

Another finding of general flow theory was very helpful here. In virtually all types of gas turbine the compressed air is decompressed, and therefore accelerated, in the diffuser system and the turbine wheel. This phenomenon is termed jet flow. If the nozzles are well-shaped the losses can be as low as 3%, giving an efficiency of 97% when converting the compression energy into flow energy. Unfortunately the specialist literature is silent about the efficiency of jet flow between turbine blades when the Reynolds numbers are necessarily very low, as in our case. To make things worse, the cross-section of the nozzles in the diffuser system and between the vanes of the turbine wheel is angular, and the nozzle ducts are also curved. Unfortunately this curvature is necessary to deflect the flow, as there would be no peripheral force at the turbine wheel without the deflection. No peripheral force equals no torque at the turbine wheel, and no torque means no shaft power for the compressor. We also have to consider unavoidable losses due to the gap between the blade tips and the housing. All in all, we are now some way away from the ideal low-loss system.

And so, back to the experiments. I used a windmill similiar to the Mallorca type as a basic pattern for the turbine wheel; this is similar to the wind-driven water pumps seen in Western films. These wheels can certainly be classified as turbines, since they are capable of converting the flow energy of the wind into mechanical energy. However, the main reason for this choice was the ease of making them at model scales. The material I used was 0.5 mm thick stainless steel sheet. The original wheel which I used in these experiments still survives. This turbine wheel was installed in a pipe with an internal diameter of 65 mm, and a peripheral gap of about 0.5 mm. A diffuser system, also made of formed stainless steel sheet, was installed in the pipe in front of the turbine wheel. The diffuser system is simply a ring of curved ducts. The curvature is designed to exert a powerful twisting motion to the flow at the outlet of the diffuser system, in the same direction of rotation as the turbine wheel. From the point of view of the spinning turbine blades it then looks as if the flow is coming exactly from the front if the turbine's rotational speed is the same as that of the twisting motion. The camber of the turbine wheel blades is designed to deflect the gas flow in the direction opposite to the direction of rotation. As a result torque is produced at the turbine blades. The use of the twisting motion substantially increases the torque at the turbine wheel in comparison with the way a windmill works. Accurate alignment of the angles and the precise camber of the blades of the diffuser system and turbine wheel are not essential. For example, the windmill starts revolving even when the airflow is offset at a slight angle. Unfortunately an accurate description of the flow processes through the diffuser system and the turbine wheel can only be expressed in terms of mathematics. Calculations show that, in the ideal case, the gas flowing out behind the turbine wheel would exhibit no twisting motion, i.e. the flow would be in the desired axial direction only. The principle is that the spiral motion built up in the diffuser system is completely negated, or removed, by the turbine wheel.

Theory also predicts that flow losses are at a minimum when the gas flow is deflected by an equal and opposite amount at the diffuser system and turbine wheel.

Back to my experimental rig: the front part of the pipe formed a gas-heated combustion chamber. In place of the compressor, a vacuum cleaner fan was used to produce the compressed air. In order to place a load on the turbine wheel, a small propeller was mounted on the free end of the turbine shaft outside the pipe. The rotational speed / power graph of the propeller was established beforehand, using an electric motor of known efficiency. The rotational speed of the propeller could now be measured, and I was then able to determine the shaft power of the turbine by comparing the measured figure with the known values for the electric motor power system. It was also possible to measure pressure, temperature and air throughflow through the combustion chamber, so that I could determine the turbine's power output. The ratio of shaft power to turbine power now produced the crucial figure I wanted – namely the efficiency of the turbine stage. The measurements showed a value of 75%. By heating the combustion chamber I could plot its efficiency relative to temperature. No measurable difference was found.

The results of these measurements were so amazing that I was forced to doubt their accuracy. The next step was to construct a turbine wheel whose blade shape and camber were closer to those of a full-size gas turbine. In spite of the enormous effort involved in making this wheel, the measured results were not significantly better than those from the initial experiment. This turbine wheel was later used in the first complete experimental gas turbine to run autonomously. As expected, this painstakingly produced wheel was not cable of withstanding high rotational speeds, and was only used for basic experimental purposes. To sum up, for the requirements of a model turbo-jet it was clearly permissible to design a turbine wheel whose compromise was biassed in favour of ease of manufacture. "Correctly" shaped wheels necessarily have greater mass. As a result, a serious problem with heat transfer arises, exacerbated by the small dimensions of the turbo-jet. Heat is conducted to the shaft and the bearings, which then have to be cooled. For example, if the massive radial turbine wheel of a turbo-charger is used, then the cooling problem can only be solved by a complex oil lubrication system. Of course, when used in a car engine the lubrication system is already present. For a model turbo-jet it would have to be developed specially.

As the drawings and photographs show, the turbine wheel employed in the FD 3/64 is very similar to that used in the initial experiments. The building instructions provide full details of how to construct the wheel. But before you skip straight to that point, I strongly recommend that you read the rest of the theoretical section. Unless, that is, you are intent on inventing your own turbo-jet without my assistance ...

A little more information on the geometry of the turbine wheel, as established as a result of my experiments. The blade length is around 1/6 of the outside diameter. Reducing blade length does not appear to reduce efficiency, but for a given wheel diameter and rotational speed the air throughput is lower, and thus lesss thrust is produced. If the blade length is increased, the blade becomes heavier relative to the carrier section of the disc. The wheel's strength at high rotational speeds is reduced, and the danger of vibration fractures in the blades increases. At this point I should mention one very important point relating to the strength of the wheel, although I will offer no mathematical / physical proof: this is the mechanical connection between the wheel and the shaft. It is vital that the wheel should not be bored through. The rotational speed strength of a bored disc is only half that of a plain disc. If a turbine wheel has to be bored in the centre, this part takes the form of a thick hub.

The division of the wheel, i.e. the number of blades, has been left at 17 in all experiments to date. A higher number of blades is certainly less efficient, and is also harder to make. A smaller number involves larger blades which do not keep their shape well. In the final analysis this just means making a new wheel.

Profiling the blades is only necessary in so far as it serves to increase strength. The usual profiling of the blades of full-size turbine wheels is designed to reduce impact losses in the flow as the load changes, when mismatching inevitably occurs. According to the laws of fluid theory profiled surfaces are only effective at relatively high Reynolds numbers. Since the size of the turbine blades is so small and the temperature so high, the Reynolds numbers are much lower even than those of a lightweight free-flight model. The influence of temperature on Reynolds numbers is as follows: for a given set of conditions the value at 550° C falls to around 1/6 of that at room remperature (20° C). The unavoidable gap between the turbine blades and the housing may be up to 5% of the blade length without a significant reduction in efficiency. This means that centration in the housing does not have to be ultra-precise, which in turn minimises the manufacturing problems.

The final part of the rotor is the shaft, which transmits the turbine wheel's torque to the compressor wheel. If this were the only criterion the shaft could be made very thin, for the torque is comparatively low, even though rotational speeds are very high. Transmitted shaft power is proportional to the product of torque and rotational speed. In contrast to a piston engine, no bending forces occur at the point where the turbine wheel is attached to the shaft. Nevertheless the turbine shaft must be very stiff. The reason for this is the danger of resonance oscillation at high rotational speeds. To avoid this problem it is necessary to design the entire rotor system in such a way that dangerous resonance oscillation does not occur. If this stricture is ignored, and the shaft starts to oscillate at a rotational speed of around 60,000 rpm, the effect on the rotor is devastating.

It is not possible to avoid resonance oscillation simply by balancing the rotor very accurately. The bending which occurs as a result of gravity is minimal but physically unavoidable, and is sufficient to excite oscillation when resonance occurs. Increasing shaft stiffness by the use of specially heat-treated or hardened steel is virtually useless in warding off oscillation. When resonance occurs the shaft simply breaks instead of bending. The safest method of avoiding resonance oscillation is to ensure that the rotor's resonant frequency is at least 20% higher than its maximum rotational speed.

The method of calculating resonant frequency is stated at (1) (See Chapter 11). The resonant frequency of the shaft currently in use, calculated using this method, is over 100,000 rpm. During the development phase of the "FD 2" I used a much thinner shaft, but it failed at a rotational speed of around 65,000 rpm. This shaft, 8 mm in diameter and 120 mm long, was bent by 2 mm. Subsequent calculations showed that the primary bending resonance oscillation was to be expected at exactly that rotational speed.

Naturally the imbalance of the rotor and of the shaft itself must be kept as small as possible, in order to minimise the load on the bearings. For example, if your shaft exhibits an eccentricity of only 0.01 mm, then that exerts a periodic force of more than 10 N on the bearings at maximum rotational speed. This alternating load makes the entire turbo-jet vibrate, which manifests itself as a loud whistling. Inadequate balancing of the rotor is the model turbo-jet's main source of noise.

The relatively large shaft cross-section also helps to disperse the heat flowing from the hot turbine wheel. In an effort to amplify this effect I succeeded in developing a shaft design whose central portion is made of aluminium alloy. It is slightly thicker than a steel shaft of equal stiffness, and thereby has greater surface area through which to conduct the excess heat. Incidentally, it is slightly lighter than a steel shaft of otherwise identical properties. Naturally the shaft ends must be made of steel to take the bearings and the wheels.

When you are mulling over the likely problems related to gas turbines at model scale, and start to consider the rotor bearings at the inevitably high rotational speeds, it is easy to lose confidence. However, with high-speed bearings we are on well-researched territory, for there are plenty of other applications where fast-running bearings are required, so the method of solving the problems is already clear.

One possible solution is to use plain bearings with forced oil lubrication, as are used in turbochargers running at very high rotational speeds. The shaft "floats" on a film of oil in the bearing shells and makes no metal-to-metal contact with them. This solution is problem-free, and rates of wear are low. However, this system requires an oil circulation system, an oil pump and a relatively large quantity of oil. All of this is present in a car engine. My own experiments in this direction led me to the conclusion that this form of bearing consumes a substantial proportion of the the turbine wheel's shaft power at high rotational speeds and comparatively low torque levels. The heat which this process produces naturally raises the temperature of the oil, and if the quantity of oil is too small the oil may even vaporise and burn. This level of insecurity is not tolerable for a model turbo-jet which is used to power an aircraft.

A simpler solution to the bearing problem is to employ ballraces and oil mist lubrication. Since the mass of the rotor is small, the bearing forces are low, and small, lightweight bearings can be used. The bearing manufacturers' specifications state that the sizes required can be run at rotational speeds of up to 90,000 rpm using oil mist lubrication. The maximum rotational speed of the turbo-jet engine, as we will see in due course, is only 75,000 rpm. The oil mist lubrication system is fully automatic, and is wear-free. The oil is pumped by exploiting the pressure difference between the compressor and the shaft sleeve, which arises naturally. Fine mineral oil, as used for sewing machines, bicycles and other machines, has proved an excellent lubricant. Oil consumption is satisfyingly low. The overall drawing shows how the lubricating system works.

If you are keen to keep your bearings fit and well, I can recommend a special turbine oil such as Aeroshell Turbine Oil 560.

2.3 The combustion system

2.3.1. Fuels

The simplest method of heating the compressed air in front of the turbine stage is to burn fuel in the airstream at that point. For model flying the most suitable fuels are those which produce as much combustion heat as possible per kg of fuel, i.e. those which have a high specific heat of combustion. These include petrol, diesel fuel, heating oil, petroleum, kerosene, and also propane and butane gas. These fuels' specific heat of combustion is approximately the same, and covers the range 40,000 to 45,000 kJ/kg. There is no point in seeking other fuels of high energy density which are easy to handle, because they don't exist. Methanol and ethanol (methylated spirits) are much less favourable in terms of energy density, and are less suitable for this reason. Diesel fuel, which is very similar to kerosene, has the highest energy density of all the fuels mentioned above, is available at every petrol station, and is also the clear leader in terms of energy density with reference to fueltank volume. By these criteria diesel is the ideal fuel for our model turbo-jet engine. To the best of my knowledge the FD 2 jet turbine was the first comparable power plant to run on diesel fuel.

One of my successful early FD2 powered development models.

Fuels can only burn if they are first brought to the gaseous state and mixed with air. The mixture can then be ignited. As with a piston engine, it is crucially important that the mixture ratio is correct, and that the gases are thoroughly mixed.

Too lean or too rich a mixture is difficult to ignite and burns poorly or not at all. As you might expect, the easiest way of preparing a combustible mixture is to use propane and butane gas, as they are gaseous at room temperature and normal pressure. Propane in particular solves many fuel supply problems in a turbo-jet, and this was the basis for the first working jet turbine, flown by a British team more than 10 years ago. For model flying, however, the advantages of propane are outweighed in the truest sense of the word by one large, heavyweight disadvantage: namely it needs a pressure tank, and the tank's volume has to be approximately twice as great as for the same mass of diesel fuel, because of liquid propane's low density of only 0.5kg/l. To be fair, the use of propane makes the fuel pump superfluous because of the high vapour pressure in the tank, but even this has its drawback. As with all liquids, vapour pressure varies very greatly according to temperature. To ensure correct fuel metering a purpose-designed radio-controlled regulatory system has to be produced.

These technical disadvantages helped me to decide, right from the start, that my turbo-jet engine would run on diesel fuel. A gas turbine designed for diesel will usually also run on liquid

propane, but the opposite does not apply. On the other hand, propane is very convenient for development work and static testing.

When considering potential fuels petrol is an obvious candidate, as it clearly vaporises more readily than diesel fuel. For this reason it is worthwhile drawing a brief comparison between the characteristics of petrol and diesel. A petrol vapour / air mixture does not ignite until the temperature exceeds 600° C, while a diesel vapour / air mixture ignites at around 300° C. The lower ignition temperature is useful for our turbo-jet, since it helps to maintain a stable flame in the combustion chamber. As a result the regulation of the airflow in the combustion chamber becomes less critical. In practice I have found that it is slightly more difficult to start the turbo-jet on pure diesel fuel.

This has led to the compromise of adding 10 – 15% of lead-free petrol to the diesel fuel. Unfortunately the composition of diesel fuel is not the same everywhere. If you are in doubt, I recommend that you use JET A 1 or JP 4 kerosene. The composition of these fuels is more uniform.

2.3.2 Combustion chamber and vaporiser

As already stated, the liquid fuel must first be vaporised in the combustion chamber. Although this process is quite simple in principle, solving the problems associated with vaporisation consumed the largest slice of time in the entire experimental programme. This is the theory: just sufficient fuel has to be burned in the combustion chamber to heat the compressed air in front of the turbine to the permissible temperature of around 600° C. However, in the combustion zone the temperature is very much higher – around 1700° C, so only a small proportion of the air is ducted into the actual combustion zone. The large residue first cools the walls of the combustion chamber. The pre-heated cooling air is then mixed in the combustion chamber with the very hot gases from the combustion zone. The result is a medium-high temperature at the outlet of the combustion chamber, which is what we want. This process is as old as the gas turbine itself. Nevertheless, there are special problems to be overcome when we are working at model scales:

1. The flame in the combustion chamber must be stable over the whole speed range, including the transitional periods, whether running on the ground or in flight.

2. The medium-high temperature which we seek needs to be as even as possible over the entire cross-section of the combustion chamber outlet. In an ideal world the temperature distribution would be arranged to fall off slightly at the outer edge and in the area of the blade root at the diffuser system and the turbine wheel. Irregular temperature distribution inevitably produces more or less pronounced "hot spots" in the region of the turbine stage. Hot spots are areas where the temperature is excessive to a serious degree, and they show up as areas of the engine which glow more or less brightly. These hot areas, especially in the outer skin, produce stress and may distort the housing. In the worst case the turbine wheel fouls the outer skin.

This problem proved to be a very tough nut to crack, and took a long time and much experimenting to solve.

3. The fuel supplied should be burned as completely as possible. The combustion chamber is divided into what we term the primary zone, i.e. the area where combustion takes place, and the secondary zone, in which the hot gases are mixed with the unburned cooling air. The division of the chamber is determined by the position and size of the openings in the combustion chamber walls. Unfortunately it is not possible to take a full-size combustion chamber as pattern and simply reduce it to scale. For a given air throughput, the larger the combustion chamber and the higher the pressure (and thus the gas density) in it, the easier it seems to get the division right. The best possible shape for a combustion chamber is annular (ring-shaped), both in technical terms and also in terms of efficient exploitation of space. This form of chamber is used in virtually all professionally produced miniature gas turbines. However, these combustion chambers are still much too large for use in a model turbo-jet. The world's first jet turbine, developed by Dr. Papst von Ohain, also featured this type of combustion chamber.

As the combustion chamber size grows, so also do its weight and the weight of the surrounding housing. If we want to achieve a good thrust : weight ratio and keep the structural volume of the whole turbo-jet small, all we can do is carry out innumerable experiments aimed at minimising the combustion chamber volume.

Just as important as the combustion chamber is the design of the fuel system. Here again the initial designs of Dr. von Ohain and Sir Frank Whittle, in which the principle of vaporisation was applied, formed a useful basis. The alternative method of rendering the fuel combustible is to use atomiser jets, as employed in most full-size gas turbines, but this appears to be an unlikely prospect given the dimensions of the model turbo-jet. A relatively high pressure of around 10 atmospheres is required to operate the jets, and controllable miniature atomiser jets have yet to be invented.

The vaporiser is nothing more than a system of heat exchanger tubes located inside the combustion chamber. The fuel is pushed through the pipes by the fuel metering pump. The flow of hot gas heats the fuel, which vaporises before entering the combustion zone in a gaseous form through several openings. The ideal dimensions for the vaporiser can only be found by systematic experiment. If the vaporiser is not effective enough, part of the fuel reaches the combustion chamber in liquid form. This results in very irregular flame formation, and the turbine spits. If the vaporiser gets too hot the fuel may crack inside the vaporiser tubes, i.e. the hydrocarbons partially break down and produce carbon, with the result that the whole system cokes up. The outflow speed from the vaporiser jets must be high enough to ensure that the air and fuel vapour mix together thoroughly in the primary zone.

To start a turbo-jet the temperature of the vaporiser must first be raised to the operating level. To achieve this it is necessary to duct gas fuel to the combustion chamber instead of diesel fuel. Propane or propane-butane gas works well for this purpose. Heating only takes a few seconds. The gas container is not part of the airborne system. Full details of the starting procedure are included in the operating instructions.

Stainless steel has proved to be a good constructional material for the combustion chamber and vaporiser. The design and method described in the building instructions form only one of many alternatives. However, in view of the many unsuccessful experiments during the development period, which I will not describe in detail, I do urge the experimentally minded modeller not to alter the geometry of the combustion chamber.

2.4 Temperature problems

As with all internal combustion engines it is essential to pay due attention to the effects of the inevitable high temperatures which affect virtually every individual component. High temperatures have two effects: most bodies expand to a greater or lesser extent when heated, and their strength is always severely reduced as temperature rises.

The heat expansion of a component can be calculated if you know the thermal expansion coefficient of the material and the change in temperature. However, it is hardly possible to calculate accurately the temperatures which occur in each state of operation. What makes things more difficult is the fact that the temperature change does not proceed simultaneously in components which are connected to each other, because heat transfer within a component may vary widely according to its efficiency as a conductor and disperser of heat. In an unfavourable case these effects may cause severe stresses within a component, with resultant fractures or permanent distortion.

Even so, we can carry out approximate calculations to simulate certain extreme conditions. Please refer to the section on Design Calculations (Chapter 4.2) for further details. When the turbo-jet is running, you will discover how accurate and reliable your calculations were. The basic rule is to design the engine in such a way that components which are inevitably subjected to major temperature fluctuations are able to expand unhindered. The FD 3/64 turbo-jet described in the building instructions has been designed in this way, and is a result of both theory and practice. For this reason it is dangerous to make alterations to the design and material selection without considering the temperature problems.

To help clarify the situation we will discuss the problems surrounding the turbine wheel. The wheel must be able to rotate freely whatever the engine's condition. If the turbine blades foul the housing the effect is similar to a piston seizure in a piston engine. Because of the turbine blades' high peripheral speed, it is not possible to achieve an oil-lubricated sliding fit, as between a piston and cylinder. The only option then is to leave a gap between the blades and the housing. However, the gases will obviously flow through the gap if possible, so the clearance must not be too wide, otherwise the gas losses will be so severe that there will be insufficient energy to drive the turbine wheel. When the turbine wheel heats up to running temperature it expands, and the gap closes. You might counter this by remarking that the housing also heats up and expands, so that the gap is maintained. This is basically correct, but there is a catch. When the engine starts running, or when its running speed changes, the internal temperatures start to change, but the various parts do not heat up at the same rate. The turbine blades and also the vanes of the diffuser system encounter a hot gas flow on both sides, and therefore reach operating temperature more quickly than does the housing. At the same time the housing is surrounded by the colder ambient air and therefore fails to reach the same temperature as the turbine blades even when temperatures have stabilised. Inevitably the gap then becomes narrower. If the clearance is too tight to start with, it is inevitable that the turbine will suddenly jam and wreck itself. If you succeed in cooling the turbine disc, the change in diameter will be less drastic. If, on the other hand, you stay on the safe side and leave the gap too wide, the result will be overheating and a non-functional engine due to excessive gap losses.

As already stated, the whole engine is subjected to varying degrees of temperature change, and it is an unfortunate fact that the engine's components do not heat up symmetrically over the course of one rotation, as they would in an ideal world. This unavoidably leads to temporary distortions, which have the effect of altering the centration of the turbine in the housing, i.e. the gap closes on one side. The main cause of this effect is the difficulty of achieving radially symmetrical combustion in the combustion chamber.

I was only able to solve the problem of the optimum gap width with some degree of certainty after a long series of experiments. I can now state confidently that a good starting point for the gap width between turbine blades and the outer skin is 0.6% of the turbine wheel diameter. For a turbine diameter of 64 mm this can be rounded off to 0.4 mm.

As with the turbine blades, the diffuser vanes also change length substantially compared with the outer mantle. We can solve this problem in a similar way too, by including a gap between the vanes and the inner central housing. Since these parts do not rotate, the gap width can be kept down to about 0.2% of the turbine wheel diameter. A slight distortion of the diffuser vanes due to overheating does not affect the turbine's running characteristics.

Now we shall consider the problem of strength reduction at high temperatures, again taking the turbine wheel as an example. There is no doubt that this component, next to the compressor wheel, is most highly stressed by centrifugal force, as a result of its high rotational speed. In the case of the compressor wheel we can ignore the temperature problem, since the temperature rise due to the heat of compression is only about 30°, as the pressure is so low. On the other hand, the turbine blades are surrounded on all sides by a constant flow of hot working gas, and virtually assume the same temperature, which may be higher than 600° C. In modern high-performance gas turbines the blades are cooled internally, but this is not feasible for us. Equally, we cannot work the special high-temperature alloys using our equipment; even if we could ever get hold of them.

The strength of all metals and alloys declines more or less steeply as temperature rises. However, when we consider that the working temperature of a model turbo-jet is only 600° C, then things do not look quite so hopeless. As discussed in Chapter 4, we can expect more than adequate thrust from a model jet engine with an operating temperature at the turbine stage of 600° C. Ordinary steel and tool steel cannot cope with this. Nickel-chrome steel, however, can withstand a temperature of 600° C at the turbine blades. The strength of this material at this temperature is around three times higher than that of steel. On the other hand, it is possible to cool the turbine disc itself relatively effectively, which helps to maintain its strength well. For example, the material's strength is almost doubled if the temperature is reduced from 600 to 500° C.

We must also take into account at the design stage the inevitable size changes in the ballrace sleeves due to changes in temperature. Because of their small diameter and the narrow temperature variation these changes are extremely small in absolute terms, but the tolerance in the fits is also much tighter. It is vitally important that no thermal distortion can occur between the shaft sleeve and the shaft as a result of changes in relative length. If the ballraces are placed under severe axial stress they will inevitably overheat. The bearings are then ruined, and the engine in turn suffers the same fate. The simplest remedy for this is to ensure that the ballrace sleeves at the turbine end are a free-moving sliding fit.

However, the ballraces themselves also have limits in their ability to withstand thermal load. The balls and bearing surfaces are made of hardened steel. At temperatures above 260° C this material suffers a marked loss of strength. For this reason it is vital to include measures aimed at reducing heat transfer, and simultaneously to cool the bearings. The principle employed in the FD 3/64 is shown clearly in the simplified cross- sectional drawing of the turbo-jet engine.

We can expect the greatest change in relative size within the engine to occur between the length of the combustion chamber and the housing, and this variation may amount to around 1 mm. In fact, this presents no problem provided that the combustion chamber is not prevented from assuming a shape appropriate to its temperature. The components which locate the combustion chamber must be designed as spring elements, and must be sufficiently flexible. In full-size gas turbines the combustion chamber takes the form of segments which can slide over each other, but this is not necessary for our application.

2.5 Cooling

The air passing through the compressor is available to us as a means of cooling the engine. The component which most urgently needs cooling is undoubtedly the combustion chamber. Around 3/4 of the total airflow passes round the combustion chamber before the cooling airstream is mixed with the very hot gases from the combustion zone inside the combustion chamber. It makes no difference whether the temperature of the gas is raised to the correct operating temperature when flowing round the combustion chamber, or whether it waits until it is mixed with the hot gas – this does not affect the running of the engine. This is true as long as no air is lost and there is no severe flow resistance which has to be overcome. The cooling of the combustion chamber has no significant influence on the overall efficiency of the rotor. We have to accept a loss of pressure of around 5% as a result of the inevitable flow resistance where the air passes into the combustion chamber. As has already been discussed, the bearings must also be cooled, and around 3 to 4% of the total airflow is required for this. This airstream simultaneously transports the oil mist through the bearings. The effect of the pressure loss as air enters the combustion chamber, and of the sacrifice of air to cool the bearings, is as if the compressor stage were working at slightly lower efficiency than as measured on the test bench. Naturally, the only method of determining the correct size of the cooling air ducts is to carry out experiments. Once that is settled, the cooling system works perfectly.

A common question is this: how hot do the external surfaces of the jet turbine become? The hottest part of the engine is the housing in the area of the diffuser vanes, turbine wheel and outlet nozzle, and the outside temperature is around 450 to 500° C. Additional cooling to improve the engine's running characteristics is not required. If you wish to cool this area, you can enclose it in a "cooling duct", which is straightforward and incurs no loss of thrust at all. The principle is shown clearly in the diagram.

The running temperature of the other parts of the housing is lower than 100° C. You can even touch the front part of the housing while the engine is running, with no danger of burning yourself.

Chapter 3

The Jet Engine and the Model

3.1 The essential differences between jet turbine propulsion and propeller propulsion

As yet there has been no chance to build up a great fund of experience with turbo-jet propelled model aircraft. Nevertheless, we can do a lot better than simple blind testing. A little applied physics can be a very reliable friend in assessing the possibilities and likely limits of this relatively new form of model aircraft power plant. For a given model it is easier to calculate the effectiveness of a turbo-jet engine in the different phases of flight than that of a propeller-driven or impeller-driven (ducted fan) model. The reason for this is that the turbo-jet engine's thrust, as measured on the testbench, can reliably be assumed to be constant up to very high airspeeds in the model aircraft. In contrast, the thrust produced by a propeller or impeller varies in a very complex manner with relation to airspeed. There is therefore little point in comparing these completely different types of power plant based on static thrust measurements, and using the figures to help decide which is the more powerful. The same logic applies to comparisons between piston engines.

The propeller converts the piston engine's shaft power into thrust. There is always a finite limit to this process: the product of thrust times speed can never be greater than the momentary shaft power of the motor. Speed equates to distance divided by time. Thrust is force in the direction of movement. Force times distance is work, and work divided by time is power. This can all be summed up as follows: thrust times speed is the engine's power for flight. Since there are inevitable losses involved in the propeller, the flight power must always be less than the engine's shaft power. Finally, the thrust in flight is defined as flight power divided by speed.

Using one and the same motor – piston or electric – it is possible to produce virtually any static thrust you like by using different propellers and different gearbox reduction ratios. Note that this can be done without varying the motor's shaft power. Whether the result is a practical power plant for a model aircraft, i.e. with its varying airspeed, depends very markedly on the characteristics of the propeller and the motor. To calculate the thrust – speed graph, i.e. the power curve of a propeller-based power system, it is necessary to superimpose the characteristic curve of the motor on that of the propeller. Of course, we don't have to go to all this effort just for normal model flying. It is perfectly possible to fine-tune the combination of motor, model and propeller by ear, by eye, and by instinct, provided that you do not move too far away from the manufacturer's recommendations.

Of course, the jet turbine is not a perpetuum mobile – it cannot produce energy or power from nothing. Even so, the physical explanation for its virtually constant thrust is clear enough. The primary power – the thermal energy of the fuel – is enormously high. For the FD 3/64 this figure is 80 kW at full load. As engine speed rises, the rate of energy conversion, i.e. the engine's efficiency, rises constantly. This is a fundamental characteristic which it shares with full-size turbo-jet engines. The graph shows the typical thrust / speed and power / speed curves for a propeller-based model power system and the FD 3/64 turbo-jet. The piston engine is assumed to be a 10 cc motor producing 1000 W shaft power at 12 000 rpm. The propeller is of 28 cm diameter and 18 cm pitch, and is a good match for the engine. The propeller's maximum efficiency is 75% at an advance ratio of 0.7. The following approximate formula can be used to calculate the motor's static thrust:

$$F_{st} = 0,6 \cdot \sqrt[3]{D^2 \cdot 3,14 \cdot \frac{\rho}{2} P^2}$$

Fst = Static thrust in N
D = Propeller diameter in m
P = Shaft power in W
p = Air density = 1.2 kg/m3

The result is a static thrust of 33 N.

The graph clearly shows that the propeller provides higher

Turbo-jet compared with propeller
Graph showing thrust and flight power relative to airspeed
Propeller 28/20 cm, 10 cc motor
"FD 3/63" turbo-jet

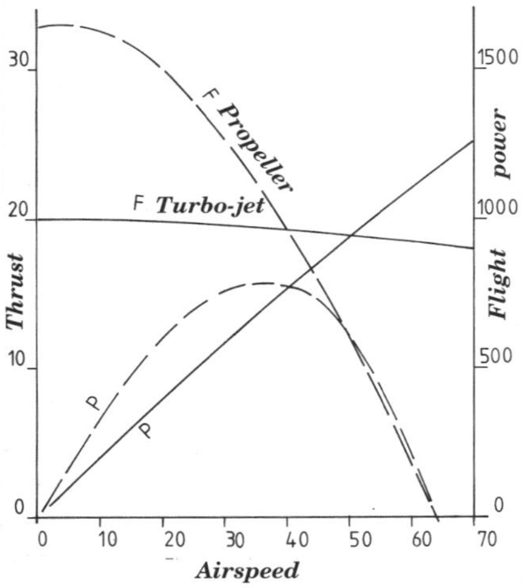

flight performance at take-off and when climbing at low speed. Above a speed of 40 m/s the turbo-jet quite clearly gives higher performance. Of course, by fitting a different propeller to the same motor the curve can be shifted, i.e. you can obtain maximum power at higher speed, although the static thrust will be reduced. However, the basic shape of the curve is the same. There are other important differences between these two types of engine which are difficult to explain. In a dive there is a limit to the propeller's rotational speed determined by the motor itself. At this point a propeller power system turns into a propeller braking system, if we assume that the motor can survive the excessive rotational speed. In practice this means that a propeller-driven model can never exceed a certain speed in a dive. In contrast, this braking effect simply does not occur with a turbo-jet engine. Even when a turbo-jet is set to what we term its idle setting, the engine still supplies thrust when the aircraft is in a dive. This has implications for your choice of model – of which more later.

The last point to remember is the torque of the piston engine. Torque effects are absent from the turbo-jet. The same applies to a ducted fan power system, since in both cases the twisting motion of the propulsive airflow is virtually balanced out by the inter-action of rotor and stator.

3.2 The forces acting on a model aircraft in typical phases of flight

With any type of power plant the basic question is always the same: what are the forces of resistance which the engine has to overcome in the various phases of flight? For example, what is the rolling resistance during a ground take-off? How does air resistance (drag) rise with speed? To what extent does drag rise in a tight turn or loop? If we are considering the engine, the question is: how much thrust must the engine produce in each situation?

3.2.1 Ground take-off

We need to know the model's rolling resistance here. This varies directly in proportion to model weight, the characteristics of the take-off strip and, of course, the quality of the undercarriage. Obviously a grass strip presents higher rolling resistance than a hard strip. My own measurements indicate a minimum rolling resistance of around 20% of the model's weight for a grass strip and a medium-sized model. On a hard surface the figure is only 5%. If we assume the model's take-off mass to be 4 kg, then the rolling resistance is:

$$0{,}2 \cdot 4 \cdot 9{,}81N \approx 8N$$

If the turbo-jet engine produces a thrust of 20 N, 12 N thrust is left to accelerate the model to lift-off speed. If the model is set up at zero degrees angle of attack on the ground, for practical purposes we can ignore air resistance until lift-off. The 12 N thrust accelerates the model at a rate of 12/4 m/s2

$$b = 3m \,/\, s^2$$

But what is the lift-off speed v? To calculate this we need to know the wing area and the maximum lift coefficient (ca) of the wing section. The relationship between lift, speed and wing area can be calculated as follows:

$$F_a = c_a \cdot A \cdot \frac{\rho}{2} v^2$$

If we assume a high-speed airfoil, then a maximum ca of 0.6 is a safe assumption. We will assume the wing area A of our model to be 0.5 m2. Air density is 1.2 kg/m3. These figures give a wing loading of 80g/dm2, or more accurately 80 N/m2. Fa corresponds to the aerodynamic lift of the model = 9.81 .4N. The formula for calculating v is as follows:

$$v = \sqrt{\frac{2 \cdot F_a}{c_a \cdot A \cdot \rho}} = \sqrt{\frac{2 \cdot 9{,}81 \cdot 4}{0{,}6 \cdot 0{,}5 \cdot 1{,}2}} = 14{,}8 m / s$$

The time t which elapses until lift-off speed is reached is:

$$t = \frac{v}{b} = \frac{14{,}8}{3} s \approx 5s$$

The last factor to be calculated for take-off is the ground-roll distance s. If we assume a constant rate of acceleration, then this formula applies:

$$s = b / 2 \cdot t^2$$

Using this formula the ground-roll s = 37.5 m.

Most model flying sites have take-off strips much longer than this. It is therefore safe to say that a thrust : weight ratio of 0.5 is adequate for ground take-off from a grass strip, provided that the wing loading is not excessive.

If you are designing a turbo-jet model aircraft it is easy to keep within the limits stated above. Semi-scale models of modern military jets are likely to present rather more problems, especially if you opt for the "glass armour plating" style of construction which is generally popular. Delta models with the mass and wing area as stated above will present no problems with the ground-roll distance.

I will discuss a number of special characteristics when describing the models.

3.2.2 Climb performance and maximum speed

Maximum speed vmax in level flight is achieved when the model's drag Wges is exactly the same as the engine's thrust. This calculation assumes that the model has not been dived to speed from a great height. Wges can be found as follows:

$$W_{ges} = v^2 \cdot \frac{\rho}{2}\left(c_{WFL} \cdot A + c_{WR} \cdot A_R + c_{WF} \cdot A_F + c_{WiFl} \cdot A_{FL}\right)$$

cw refers to the drag of wing, tail, fuselage and undercarriage. A is used for the corresponding surface areas. The area of the fuselage and undercarriage is the projected area perpendicular to the direction of flight. For a model of the size we are discussing the values are summarised in the table below.

Model component	Fl . A m2 m^2	Drag coefficient c_w	A . c_w . %
Wing	0,5	0,006	0,0018
Fuselage	0,011	0,1	0,00066
Undercarriage	0,004	0,6	0,00144

Cwi can be ignored when the aircraft is at high speed.

If we are aiming at an accurate calculation it is necessary to include the influence of Reynolds numbers. Assuming turbulent airflow around a flat plate the Reynolds number reduces at the rate of:

$$\sqrt[5]{\frac{Re_1}{Re_2}} ab$$

If speed is doubled, this equates to a reduction by a factor of 0.87. However, this effect is only approximate, and applies to the wing and tail only. Since we are only aiming at an approximate figure, we can safely ignore these finer points. If we now put these values together and assume the engine's thrust to be 20 N, then the formula looks like this:

$$v_{max} = \sqrt{\frac{20}{0{,}0039}} m / s = 71{,}6 m / s$$

The maximum level speed of our model is therefore vmax = 71.6 m/s or 258 km/hr. You will need iron nerves and perfect vision if you keep the throttle wide open. Even if engine thrust is reduced to half, the airspeed only declines by $1/\sqrt{2}$, which is still about 180 km/hr.

Next comes climb performance at a constant velocity vb. We can think of this phase of flight as pushing the model upward along an angled plane, at an angle of ascent α. In this situation the force G · sin α acts in the opposite direction to the direction of flight, as also does drag W. Speed remains unchanged as long as F = W + G · sin α, i.e. the forces are in a state of equilibrium. The rate of climb is then calculated as follows:

$$v_{steig} = v_b \cdot \sin\alpha$$

where $\sin\alpha = \frac{F - W}{G}$

and G=m·g

This can also be expressed as follows:

$$v_{steig} = v_b \cdot \frac{F - W}{G} = v_b \cdot \frac{F - W}{m \cdot g}$$

We can use the data from the preceding table to calculate drag. For the moment we can ignore the speed-dependent influence of induced drag, which is only valid if

Operational data for a model aircraft relative to thrust

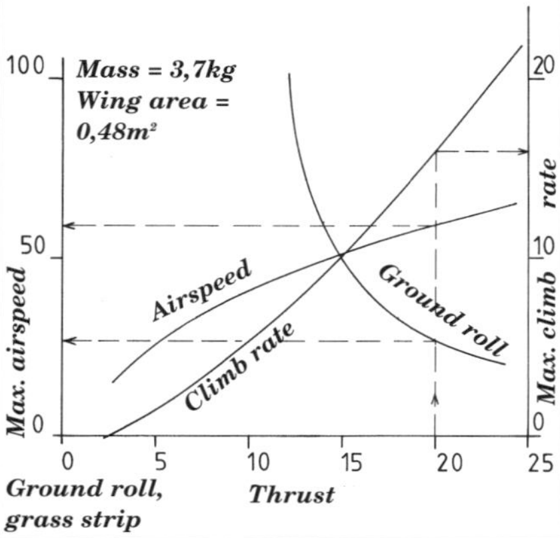

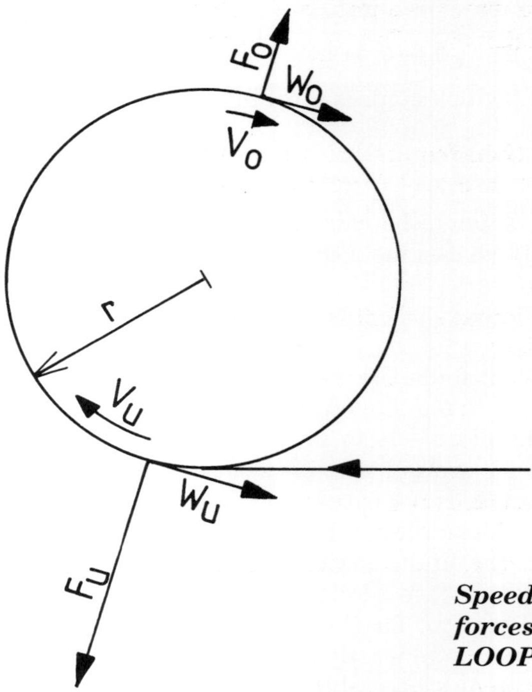

Speeds and forces during a LOOP

the model's speed is considerably higher than its lift-off speed. As is well known, induced drag declines in inverse proportion to the square of the model's speed. Since even an approximate calculation is very complex, I have presented the results in the form of a graph. This graph shows the data for my latest "Rutonius" model, and represents the potential maximum speed, minimum ground-roll distance and optimum climb speed relative to thrust. With a thrust of 20 N the model climbs at 13 m/s at an air-speed of 40 m/s. An easy way to summarise these results is this: a turbo-jet powered model which can take off reliably from a grass strip has an absolutely top-class performance once in the air. A thrust / weight ratio of 0.5 is quite sufficient for this, even for many semi-scale models.

3.2.3 A typical power manoeuvre: the loop

Can the model perform a large loop? It is quite easy to estimate the maximum effort required for this aerobatic manoeuvre. The drawing should help to clarify matters.

If we consider the energy balance we will quickly see the explanation. The model flies at an approach speed vu and therefore has the stored energy Ekin = m/2 . vu2. M is the mass of the model in kg. In reaching the top of the loop the model covers an altitude difference 2r. This requires the model to consume the potential energy Epot = m . g . 2r, regardless of the route by which the model reaches this point. We also have to take into account the losses due to drag as the model flies from Fu to Fo. Since the model's speed can be expected to decline as it climbs, drag also falls off as speed decays. A considerable amount of mathematical effort is required to obtain an accurate calculation of the energy balance involved in this manoeuvre. However, since we only need an approximate answer, we can simplify the matter by assuming that the influence of drag at the maximum speed vu remains constant over the entire course of the manoeuvre. In simple terms, we can rest assured that the situation is rather more favourable than this in practice. The engine's thrust Ft acts in exactly the opposite direction. We can reasonably consider thrust to be a constant. The effective drag force along the flight path would then be Ft – F1, and the energy balance Ex = r . (Ft – F1). In the worst case the model would have the kinetic energy Ekino=Ekinu + r · π · (Ft–F1) – 2r · m · g at the top of the loop.

If we calculate the formula and come to a value of less than zero, then the model cannot fly a loop of radius r. If the value is greater than zero we can calculate the speed vo according to the formula:

$$v_o = \sqrt{\frac{2E_{kino}}{m}}$$

So far so good, and all that remains is to calculate the drag factor. This is at its maximum at the start of the loop, i.e. when the pilot first pulls back the stick. At this point the model experiences a force which is additional to its weight, acting perpendicular to the flight path. This is due to circular acceleration:

$$b_{kreis} = \frac{v^2}{r}$$

The wing is then required to produce the lift Fa = m . (g + b). For this we need to know the lift coefficient ca:

$$c_a = \frac{2 \cdot F_a}{A \cdot v_u \cdot \rho}$$

Once we know the value of ca we can calculate the induced drag coefficient:

$$c_{wi} = \frac{{c_a}^2}{\pi \cdot \lambda}$$

π = pi
λ = wing aspect ratio

The additional induced drag can be calculated by the formula

$$W_i = {v_a}^2 \cdot A \cdot c_{wi} \cdot \frac{\rho}{2}$$

Of course, the profile drag of most airfoils also increases as ca rises. If you have access to the profile data you can find this value from the characteristic curves.

To illustrate the problem we will calculate the data for a given situation. We will stay with our model of 4 kg mass, capable of ground take-off, and fitted with an engine producing a thrust of 20 N. We will assume an arbitrary entry speed vu of 50 m/s at the start of the loop. Bearing in mind the previous chapter, this is a reasonable figure. We will aim at a loop of radius r = 40 m. The wing section is NACA 009, i.e. a typical high-speed profile. To make the matter clear, I will summarise the essential data once more, and complete the calculations:

m = 4kg
Ft = 20N
A = 0,5m²
vu = 50m/s
r = 40m
λ = 5

$$b_K = \frac{50^2}{40}\frac{m}{s^2} = 62{,}5 m/s^2$$

$$F_a = 4(9{,}81 + 62{,}5)\frac{kgm}{s^2} = 289\frac{kgm}{s^2} = 289N$$

$$c_a = \frac{2 \cdot 289}{0{,}5 \cdot 2500 \cdot 1{,}2} = 0{,}385$$

$$c_{wi} = \frac{0{,}385^2}{3{,}14 \cdot 5} = 0{,}00945$$

$$W_i = 50^2 \cdot 0.5 \cdot 0{,}00945 \cdot 0{,}6N = 7{,}1N$$

Based on this figure for Ca, the profile drag coefficient rises from approximately 0.006 to 0.01. The drag of the wing is now:

$$W_i = 0{,}01 \cdot 2500 \cdot 0.5 \cdot 0{,}6N = 7{,}5N$$

The additional drag of fuselage and undercarriage were calculated in the preceding section in order to determine maximum speed. We can assume these figures are unchanged, and at a speed of 50 m/s their drag is 5.3 N. The total drag is therefore 5.3 N + 7.5 N + 7.1 N = 19.9 N, i.e. virtually identical to the engine's thrust. Thus the overall energy balance as the model flies to the top of the loop would be:

Ekinu	=	5000Nm
2r·m·g	=	–3139Nm
wu·r	=	13Nm
Ekino	=	1874Nm

From this we can see that the model's speed vo is at least: 30.6 m/s.

At this speed the outward acceleration at the top of the loop is: 23.4 m/s2.

At this point the earth's acceleration acts in the opposite direction, towards the centre of the loop. However, this acceleration is only 9.81 m/s, which is considerably smaller. This means that slight up-elevator will still be required, even at the topmost point of the loop. This approximate calculation shows that a model with a thrust : weight ratio of only 0.5 is still capable of very passable aerobatic manoeuvres, although it is important to realise that the model's initial speed must be fairly high.

As our calculations have shown, we cannot ignore the increase in overall drag due to the considerable increase in induced drag as a result of radial acceleration. Now let us consider a delta aircraft for comparative purposes. If we assume that the delta has the same mass and the same wing area, the increase in induced drag would be roughly five times higher compared with the conventional model we have just discussed. This means that deltas lose speed extremely quickly in a fast, tight turn. Similar effects can be observed with models based on modern military jets. You will need an extremely powerful engine if you want to fly impressive aerobatic manoeuvres with this type of model. To fly the same manoeuvre with a delta you will need a thrust : weight ratio of approximately 1, i.e. 40 N thrust. At a speed of 50 m/s this equates to 2000 W of flight power. If you are aiming at this level of power using a piston engine and ducted fan, then you must also take into account the power plant's efficiency as installed in the model. If we estimate this efficiency to be around 60%, we find that we need an engine whose shaft power is 2000/06 W = 3333 W or 4.5 BHP. Note that we have not even considered the extent to which the model's drag is increased by the additional air inlets required for ducted fan operation. In contrast, a turbo-jet gets by perfectly well with an inlet opening of only 60 mm diameter.

3.3 Flying experience with turbo-jet models

3.3.1 Turbo-jet models to date

I have had five years of experience of flying turbo-jet powered models, using a total of five different types of model fitted with FD-series turbo-jets in advanced stages of development. The essential model data is summarised in the table below.

The table only reflects my own practical experience. Many model aircraft powered by FD engines have now flown successfully, and covering all of them in detail would fill a new book by itself.

Model name	Mass kg	Wing-span m	Wing area m2	Date of first flight	Engine	Thrust N	Rotational speed rpm
Elkete	3,1	1,5	0,4	10.9.89	FD2	10	63000
					FD2s	17	75000
Mirage 2000 (Delta)	4,1	1,11	0,5	21.5.90	FD3	18	70000
Turbo-Scnurf (Delta)	3,3	1,15	0,55	20.10.90	FD3/62	13	73000
					FD3/64	18	75000
Rutonius	3,8	1,77	0,48	2.8.91	FD3/64s	20	67000
F100	6,2	1,47	0.6	11.92	FD3/64s	22	72000
					FD3/64ls	30	87000

Elkete

This was a purely functional model, and was probably the first turbo-jet model to fly running on liquid fuel (diesel fuel). Although its thrust : weight ratio was only 0.32 it could take off from the ground easily from a tarmac strip.

After I had made improvements to the engine, the model was also able to take off from a grass strip. I was able to confirm the calculations relating to take-off distance (see Section 3.2.1) by studying video pictures of the take-off phase. The model's speed was not particularly remarkable, as a result of the large diameter of the engine and its high mounting. What did astonish me was the lack of pitch trim change in flight when thrust was varied, bearing in mind the engine's thrust axis, located high above the Centre of Gravity. When the throttle is opened it seems likely that the secondary stream of cold air is deflected onto the tail at a slight upward angle, which has a similar effect to applying up-elevator. This appears to compensate automatically for the turning moment around the lateral axis. Reiner Binczyk's first experimental model was similar.

Mirage 2000

This was a modified semi-scale model. With a thrust : weight ratio of less than 0.5 and a wing loading of slightly over 80 g / dm2 the model needed a ground run of around 40 m. Its climb rate and top speed were very impressive, and it could still be flown fast with thrust cut back. The small scale-sized air intakes amazed ducted fan pilots.

Turbo-Schnurf

This was also a delta, but it had a much lower take-off mass and a thinner wing section than its predecessor. Even with a static thrust of only 13 N it took off from the ground reliably. Once the model was in the air, the 13 N thrust could be reduced considerably without the model appearing to be under-powered. When the model was fitted with the more powerful FD 3/64 engine the maximum thrust of 18 N was only used for take-off, steep climbs and loops. In level flight at full throttle it was acknowledged to be the fastest model ever seen at the flying site.

Rutonius

The wing geometry of the Rutonius is conventional, with a

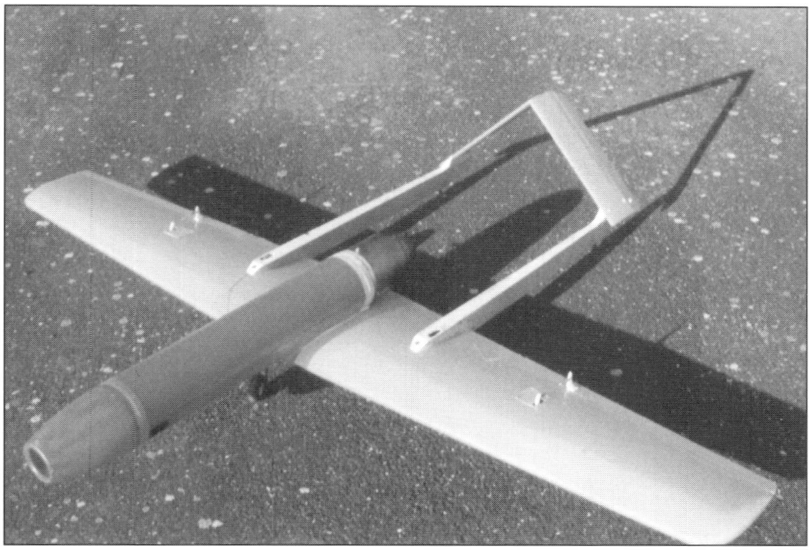

"Rutonius" after converting the tail. The repairs were necessitated by the defective tail arrangement, as mentioned earlier. The air inlets above the engine are not absolutely essential.

"Rutonius", a functional turbo-jet model for the "FD 3/64". Here again the air inlet is through the fuselage. The wheel under the fuselage is steerable and retractable. Small support wheels are recessed into the ends of the tail booms. This design of undercarriage has proved outstandingly good, but the tail arrangement, in contrast, was disastrous.

profile thickness of 15%. With its favourable thrust : weight ratio of better than 0.5 the model's flight performance is outstanding. In spite of the thick wing section it flies extremely fast and converts speed into height very effectively. It is a genuinely aerobatic model. On the landing approach the model lacks the braking effect which is characteristic of a throttled-back propeller engine. As a result it lands like a high-performance glider unless the engine is stopped.

The model's high-set T-tail exhibits the opposite tendency to that of the Elkete. When the engine is run up from idle to full throttle, the model's nose drops. When the throttle is closed, the nose rises slightly. This phenomenon seems to confirm that the pitch trim changes are due to the asymmetrical effect of the secondary airflow over the tail. In the original version the Rutonius' tail took the form of a pitched-roof inverted V-tail. In this configuration the model was almost uncontrollable, and this almost resulted in a total write-off.

The undercarriage of the Rutonius is unusual in design. It consists of a single steerable, retractable wheel in the centre of the fuselage, about 10 cm forward of the Centre of Gravity. Two small wheels at the end of the tail booms provide very good tracking on the ground. The undercarriage gives the model an angle of attack

In the interests of safety the hot exhaust gases are deflected below the tailplane through a sheet metal duct while the engine is being started. The starter air is blown in through the main inlet opening, at which time the supplementary air flaps close automatically. Once the engine has stabilised at its idle speed, the exhaust duct is no longer needed.

of about 5°. This arrangement has proved excellent over many flights, especially when taking off from a grass strip.

F 100 Super Sabre

Spectators and model flyers alike think of this machine as the archetypal "jet fighter". The full-size machine first flew in 1953, and was one of the first fighter aircraft to break the sound barrier in straight and level flight.

Many different manufacturers in all parts of the world offer kits for this model for ducted fan power. Like all models of this type its severely swept wing planform endows it with certain characteristics which are undesirable for model airspeeds. The reason for this is as follows: compared with a sports model of conventional design, a wing with a lot of sweep-back requires a much greater angle of attack relative to the air-flow to produce the same amount of lift for a given airspeed and wing loading. The usual measure of lift is the lift coefficient ca. For example, a ca of 0.8 can be assumed for a sports model at take-off, which equates to an angle of attack of about 6 – 8°. If you now turn to a model with a strongly sweptback wing and look for the same value of ca, then the angle of attack has to be around twice as great, i.e. 12 – 16°. However, such an angle of attack produces much higher drag, both of the wing and of the steeply angled fuselage. The result may be that, even though the model has sufficient speed to lift off with the help of ground effect, a powerful application of up-elevator produces the sharp increase in drag as just mentioned. This in turn brakes the model abruptly, and it drops back to the ground. You can avoid this problem by letting the model accelerate on the ground for as long as possible without pulling back, so that it's speed is higher when you ask it to climb, and it climbs safely at a lower ca value, and therefore with lower drag. The net result is that the ground roll of this type of model is much longer than for a "normal" model.

These fundamental differences apply regardless of the type of power plant. Since the landing strip at most model flying sites is of limited size, you have to come to terms with this unavoidable fact of life right at the outset when flying jet models.

To reduce this problem it is important to keep the wing loading of such models within limits, and to ensure that ground roll resistance is as low as possible. This involves careful undercarriage design, strong wheel bearings and the largest size of wheels you can fit. Of course, the thrust : weight ratio must also be adequate. In practice you will find that although maximum thrust may be required for take-off, you will hardly be able to exploit such levels of thrust at all in normal flying.

Modelling a full-size prototype like the F 100 to produce a genuine model jet fighter, i.e. one fitted with a model turbo-jet, presents a further challenge to the designer and pilot: the jet engine must be fully integrated into the fuselage. In this respect we probably have more experience with the F 100 than with any other model of its type.

I based the F 100 described here on a kit produced by the Kudelka company. The fuselage is a laminated epoxy – glassfibre moulding, while the wings and tail panels are obechi-skinned foam cores. The tailplane panels and a small part of the wing are removable to ease transport problems. Controls include ailerons, operated via two wing-mounted servos, an all-moving tailplane, steerable nosewheel, and of course variable engine power. The undercarriage is pneumatically retractable.

The following table shows the mass of the components required for the power system alone:

Engine	800 g
Pump	110 g
Pump battery	120 g
Thrust pipe	110 g
Electronic control unit	40 g
Oil tank incl. oil	30 g
Pipework, valves	30 g
Fueltank	120 g
Fuel (1 l. diesel)	340 g
Total:	2,210 g

You may like to draw up your own comparative table for an alternative power system, and compare the weights.

The model's all-up weight is 6.2 kg, and this includes fuel. Of this the airframe and radio control system account for almost exactly 4 kg. To improve the take-off characteristics it would be necessary to save weight here, without reducing the size of the model. Making the model smaller would indeed save weight, but at the same time would increase the wing loading. This in turn would result in a very high landing speed, and so the space problem would not be solved. The flying characteristics of the F 100 as presented here are as follows: with a static thrust of 22 N a 50 m length of asphalt strip is sufficient for take-off in nil wind conditions. The model looks very realistic in the air, and thrust is adequate; it is enough for slow rolls, for example. With 25 N static thrust and a very good grass strip the ground roll is about 80 m. At 30 N the take-off problems more or less vanish. Provided that the model is flown in broad curves, it can complete large loops, fast and slow rolls and impressive climb-outs with ease.

Now we'll take a look inside the model: the fueltank must, of course, be located at the model's Centre of Gravity. The engine is installed aft of the tank, with about 5 – 6 cm free space between tank and air intake. The large fuselage cross-section makes special air ducts superfluous. The engine's exhaust stream is ducted through a thrust pipe 43 cm long. An interesting feature here is that there is a gap of about 10 mm between the engine's jet outlet and the thrust pipe's inlet, which is rounded. The diameter at the thrust pipe's inlet is about 80 mm, and around 75 mm at the outlet. The thrust pipe is formed from 0.1 mm stainless steel sheet. The seam can either be riveted or spot-welded. Air is sucked in between the jet outlet and the thrust pipe inlet, and this air is mixed with the hot exhaust gases inside the thrust pipe. As a result the temperature of the walls of the thrust pipe stays below 350° C. The exhaust temperature at the end of the thrust pipe in the centre was measured at 380° C. By carefully adjusting the gap between the jet and the thrust pipe it is even possible to achieve an increase in thrust of 1 – 2 N. The transitional area between the end of the fuselage and the thrust pipe is made of nickel-chrome steel sheet. The outlet diameter of this nozzle is about 5 mm larger than the outlet diameter of the thrust pipe. The thrust pipe ends about 3 mm inside the nozzle. With this configuration there is no detectable loss of thrust, and additional cooling air is sucked in between the nozzle and the end of the thrust pipe.

The rear part of the fuselage is protected from heat by a layer of thin, self-adhesive aluminium foil applied to the inside of the fuselage; this is quite sufficient. The protected area extends from the front end of the engine to the tail end of the fuselage.

When the aircraft is at rest or at moderate speed there is a slight vacuum inside the fuselage, with the result that the engine access flap is visibly sucked inward. But even at quite moderate speed the internal pressure predominates, and if the flap is not well secured it simply gets ripped off. The low internal pressure in the take-off phase makes it necessary to seal off the undercarriage wells on the inside, otherwise there is a danger that dirt and grit from the undercarriage will be sucked up and into the engine.

Increased care is required when starting the engine. The exact procedure is described in detail in Chapters 9 and 10. You should be aware of one special danger with an enclosed engine: when filling the fueltank it is essential to ensure that the engine does not get flooded with fuel. If you overlook this, liquid fuel will inevitably be expelled from the engine when the starter is applied. The excess fuel then runs around the inside of the fuselage and may catch alight. In such cases the only hope for the model is a fast response with the fire extinguisher, which must be kept close to hand. I recommend a CO2-filled extinguisher, as this leaves no residue in the area of the fire. Of course, if you have a fire it is absolutely essential that you check the whole model for fire damage (scorched cables etc.), even if no damage is immediately obvious. With an installation similar to that described here, I do not recommend external ignition, i.e. with a flame held at the end of the thrust pipe. The danger here is that the supplementary gas ignites in the thrust pipe only, and the flame stays there. After a short period this leads to overheating in the thrust pipe area, which may damage the fuselage skin in spite of the aluminium screening. It is much safer to employ glowplug ignition in this situation. If your engine is not fitted with a glowplug socket, an electric gas igniter can be slipped into the gap between jet and thrust pipe, and the engine ignited there. But please be careful, all the same. Keep the fire extinguisher handy at all times.

The remaining components – the pump, batteries, electronic control unit etc. can be installed in any location, and can be fitted in the front part of the fuselage to help the CG situation. With a fuselage of such large cross-section the flow speed is relatively low even when the engine is at full throttle, so that no significant flow losses occur. The only area where the surfaces should be smooth to aid airflow is the fuselage air intake.

3.3.2 Characteristics of turbo-jet models

Obviously the model's tail surfaces must not lie directly in line with the engine's primary exhaust flow. As a result the rudder and elevator have effect at all at low speed on the ground. A steerable undercarriage, with plenty of power on the steered wheel, is therefore essential to provide controllable taxiing behaviour on the ground, and especially for the take-off. A tricycle undercarriage ensures stable, straight tracking. Since propeller clearance is not necessary, the undercarriage legs can be kept short. It is not necessary to set up the undercarriage so that the model has a positive angle of attack on the ground; the "Mirage", with a take-off mass of 4.1, had a slightly negative angle of attack when rolling, but took off without any problems.

Semi-exposed engine installation

The engine has to suck in all the air it requires through the fuselage. The internal fuselage fittings appears to have no effect on the air supply. All that is needed is an air intake of at least 60 mm diameter with rounded inlet edges, and a short diffuser increasing in diameter to about 70 mm. It proved unnecessary to install a special duct from the diffuser to the engine's inlet nozzle. The unobstructed cross-sectional area of the fuselage at its narrowest point, after deducting the restricting effect of the internal fittings, should be at least twice as large as the cross-section of the inlet diffuser. The turbulation of the airflow caused by the internal fuselage fittings causes no problems. The air throughput of a turbo-jet engine is much less than for a comparable ducted fan installation, and that is why the smaller flow areas are adequate.

If the engine's exhaust nozzle projects out of the fuselage you can ignite the mixture with a match or gas lighter.

Centre of Gravity position

In all our models the fueltank was installed at the model's Centre of Gravity. The engine then has to be installed aft of the tank, unless you favour the arrangement used on the first experimental model, the "Elkete". With deltas there is usually sufficient fuselage length forward of the Centre of Gravity to achieve correct balance without resorting to ballast, if the batteries, receiver and pump are located in the front part of the fuselage. On the last model we needed only 40 g of lead in the fuselage nose to trim the model correctly. The method of determining the Centre of Gravity for a turbo-jet model is the same as for conventionally propelled models.

3.4 The turbo-jet engine in flight

As speed increases, air pressure acts as an additional compressor stage for which no turbine power is required. At the same time the airflow causes a slight fall in air pressure at the tail end of the engine. The result of these two effects is that the compressor appears to work more efficiently in the air than on the test bench, as air throughput rises. However, for a given quantity of fuel the temperature in the combustion chamber falls. This may result in the vaporiser ceasing to work satisfactorily, and in the worst case the flame in the combustion chamber may be blown out. However, fuel is still being fed to the combustion chamber, where it vaporises on the hot walls. This phenomenon is immediately visible as a white cloud of smoke, and the engine stops. Alternatively the fuel may be ignited by the hot parts of the engine, because the compressor immediately stops working when combustion is interrupted. If this should happen, the engine spits at irregular intervals. The problems of blowing out and spitting occurred with the first flight-capable version of the FD 2, but it proved possible to eliminate them completely by improving the design of the combustion chamber and the vaporiser. Another obvious problem arises when the vaporiser temperature is too low. In this case the engine refuses to pick up speed from a moderate throttle setting when the model is flying. The cause is that a large proportion of the fuel is simply not being burnt. Once again careful testing and tuning of the length of the vaporiser produced a remedy. These teething troubles are absent from the version shown in the drawings. The thrust of the FD 3/64 can be controlled smoothly in flight from full load to idle and back in any flight situation. Even pushing the model into a steep dive from a great height with the engine at the minimum idle setting will not blow the flame out. Nevertheless, for safety reasons you should always avoid diving at full throttle from a great height. Since the thrust of the engine is virtually undiminished, the model's speed can very quickly rise to the point where the airframe breaks up. Nowadays it is a routine matter to land the model with the engine idling, then taxi back to the pits under engine power; just like with a correctly adjusted piston engine.

Standard fuel tanks with a clunk weight have proved fine for use in turbo-jet models, although it is important that all pipelines from the tank to the pump are made of petrol-resistant, diesel-resistant rubber. Silicone tubing is completely unsuitable for petrol, diesel or any mixture of the two. In gusty weather it is sometimes impossible to avoid air getting into the fuel feed line. This will result in the engine faltering, but it won't stop unless the interruption in the fuel supply lasts longer than about a second. A very effective method of preventing this problem is to use a felt-covered clunk weight. This ensures that virtually the last drop of fuel is sucked out.

3.5 Noise

There are two different sources of noise in a model turbo-jet. The first is the sound of the rotating engine components, which manifests itself as a whistle. The volume of this sound depends entirely on the accuracy with which the engine is balanced. At high rotational speeds imbalance can be caused by a very small mass. For example, I cut a small pad of textile adhesive tape, about 5 x 5 mm in size, and applied it to the inside of the air inlet opening of the compressor wheel. When the model was in flight the whistle suddenly became loud, and I shut the engine down in the interests of safety. After the landing I inspected the engine carefully, but could find nothing amiss. The only change was the small balance mass. Test runs on the ground showed that the effect could be repeated at will. This proved that the source of the whistling sound was undoubtedly mechanical vibration as a result of imbalance. A rotational speed of 60,000 rpm equates to a whistle at about 1000 Hz.

The radiation pattern of the whistling sound is strongly directional, and is at its strongest in a direction perpendicular to the direction of flight. If you look directly at the tail end of the engine, the whistle is not audible on the ground at about 20 m range. From the opposite direction you hear the characteristic sound of the flow of the exhaust gases. The quality of sound is very similar to that of a large jet engine, but much, much quieter. In a model turbo-jet the speed of the exhaust gases is about 150 to 200 m/s, which is very much higher than the airflow from a conventional propeller, but still way below the speed of sound. The familiar thunderous noise of full-size jet engines, especially military jets, is due to two factors: on the one hand the engines' power, which is many thousands of times higher, and on the other the speed of the exhaust gas flow, which is above the speed of sound.

Noise measurements with piston-engined models are usually recorded at a distance of 7 m, perpendicular to the axis of the engine. Using the same procedure with a turbo-jet we recorded a value of 75 dBA. The engine in question was not even perfectly balanced. Our flying club colleagues, who have spent much time and effort in reducing the noise of their internal combustion engines, consider the model turbo-jet to be a very quiet source of power for model aircraft.

During the first special event for turbo-jet models – the Ohain/Whittle Trophy, held in Nordheim – the background music, which was broadcast via the loudspeakers at quite normal levels, was switched off during the demonstration flights, as the music prevented spectators hearing the models properly in flight.

I ought to mention a very peculiar acoustic phenomenon here which occurs on hot, calm Summer days. A sound source close to the ground, e.g. a model aircraft still at low altitude after taking off, becomes almost inaudible at a range of not much more than 100 m, with the result that you can easily come to the conclusion that the engine has stopped. The reason for this lies in differential air temperatures. Air temperature can be markedly warmer at ground level before a thermal bubble starts to rise. The result is that the sound of the engine is deflected upwards, and it seems much quieter at ground level.

3.6 Model recommendations

There is one basic requirement for any model aircraft fitted with a turbo-jet engine, and it should hardly need stating here: the model requires a competent pilot, confident in his ability to fly high-speed model aircraft powered by propeller engines. Purpose-designed kits, almost-ready-to-fly models and plans for fully developed and tested turbo-jet models are few and far between at present. This is one area in which much development work remains to be done.

We have not encountered any fundamental problems in the flight characteristics of our turbo-jet models compared with conventional propeller models, and we do not expect to find any. For initial test flying the safest route to a positive experience with a new form of propulsion is to convert a standard fixed-wing model such as the "Elkete". If you opt for a low-drag design with conventional wings, such as the "Rutonius", then we strongly recommend airbrakes or split flaps, otherwise you will have trouble landing the model accurately with the engine throttled back. As we all know, power flyers suffer symptoms of stress when they have to land a model with the motor stopped.

Deltas fly well and are also easy to land, thanks to their excellent stalling characteristics. On the landing approach the drag coefficient rises so steeply that you may even have to open the throttle slightly to keep the speed up. However, even a delta's docility at the stall has its limits. For example, if the engine fails it is possible to pull the model's nose up to the point where the control surfaces no longer encounter an airflow, and the model is then uncontrollable. The only possible escape route from this situation is thrust – but there isn't any! If you are lucky, the model will descend to the ground at a manageable glide angle, and the final bump will be relatively gentle. The Centre of Gravity of a low-wing delta lies above the wing. If the wing is stalled at a very great angle of attack, the centre of pressure may shift behind the model's neutral point. At that moment the aircraft becomes completely unstable, and since its control surfaces have no effect, the inevitable result is disaster. The implication is clear: you need experience with deltas before you fit a turbo-jet in one.

'Mr. Turbine', Kurt Schreckling, readies his F100 Super Sabre for another flight at Nitro-days Punitz, in Austria. This model was originally powered by an FD2, but now has an FD3/67LS installed.

Berndt Binczyk's own-design Douglas D558 Skystreak makes a great platform for all the FD3 series turbojets, and flies extremely well with the 67LS installed.

Chapter 4

Designing a Model Turbo-jet – The Calculations

The formulae explained in this chapter are developed from the theory of thermo-dynamics, and their use does call for a certain level of experience in applied mathematics. If all you want to do is apply the results, it is not necessary to grasp all the theoretical details in terms of the mathematics involved. For example, if you can easily solve a problem such as [formula], then to my mind you have sufficient understanding of maths to be able to follow the design calculations for a model turbo-jet. If you just want to construct an engine following the building instructions, even this level of mathematical understanding is not necessary. Nevertheless I do recommend that all of you should read this theoretical chapter at least once, even if you don't find everything comprehensible.

The design calculations are based fundamentally on international SI units, i.e. using kg, s, m and the units derived from them. You may find it unusual to see the outlet area or the cross-sectional area of a hole defined as m2, but at least this means that we don't have to mess about with conversion factors. The various units of measurement are explained at the appropriate point. This avoids the tedium of searching through reams of tables showing formulae and dimensions, which are not always clear in any case, since the Greek and Latin alphabets are not limitless in scope. The formulae and calculation processes I use are based fundamentally on Bohl [1], Dietzel [2] and Kalide [3]. (See Chapter 11)

4.1 Speed triangles and speed plans

In the specialist literature the speed of gas flows and of rotating blades is indicated by speed triangles drawn to scale, showing both magnitude and direction. A uniform set of abbreviations is now standard:

u = peripheral speed of the blade element in question, i.e. the speed in the direction of rotation.

c = speed in the static diffuser system, housing or nozzle area. This is also named absolute speed. It is the speed which an outside observer would measure.

w = speed with reference to the rotating blades, also known as relative speed. If a measuring system could be attached to the blades, this is the speed which the system would record at that point.

The salient points at which it is useful to know the speeds include the entry and exit point at the blades, and the entry point into the subsequent stage, if present. The speed triangles at the entry and exit points of a blade element can then be added together geometrically to form a speed plan. The next section shows how this method can be applied usefully.

4.2 Designing the turbine stage

I will start at the rear end, i.e. at the outlet of the turbine stage. This is quite deliberate, since the most important part of any turbo-jet engine is the exhaust exit speed and thus the thrust. The first question to be addressed is this: how to design a turbine stage to produce the required thrust, taking into account maximum temperature and peripheral speed. The design of the compressor stage is calculated to match the design of the turbine stage. Maximum possible economy, i.e. lowest possible fuel consumption, is not top priority for the relatively short flight times of a model aircraft.

We will restrict ourselves to an axial turbine stage. In an axial turbine the working gas flows into the blades in the axial direction and exits again in the same direction. It is a characteristic of all turbine wheels of this type that the blades are short compared with the overall wheel diameter. As a result all we need to do is consider blade speed at an average diameter dm. The sketch shows a developed view of the turbine wheel blades at the plane of the drawing, together with the corresponding stationary diffuser system. The two cross-sections are separated in the axial direction to show the speeds more clearly.

In an axial stage the peripheral speed u is the same at the inlet and outlet of the turbine blades, measured at the average diameter dm. We will consider the conditions relating to flow and movement from the outlet side, i.e. behind the turbine blades. After all, our primary concern is the final result, namely the thrust jet. This is represented by the pointer c2. Provided that c2 points directly in the axial direction of the turbine, we can consider it as the thrust vector. This point is defined at the start of the design process. c2 and the peripheral speed u of the turbine blades then form a right-angled triangle. The third side of this triangle is the relative speed w2 of the exhaust gas, as measured at the exit point of the turbine blades. This completes what we can term the outlet triangle. If the turbine blades are sufficiently close together, the exhaust gas follows the camber of the blades, so that the angle B2 between w2 and u is virtually the same as the angle at the outlet edge of the turbine blades. This formula then applies:

$$c_2 = u \cdot \tan\beta_2$$

We can now define the ratio of c2 to u by selecting the blade angle B2, and at the same time of w2 to u, and any value is theoretically possible. It remains to be seen to what extent this is possible in practice.

Of course, we also have to ensure that the essential conditions which allow this operational situation are physically present.

To give you an idea of how the theory works out in practice, we will carry out a mathematical estimate of the maximum thrust of the FD 3/64 turbo-jet engine. To simplify matters we will ignore the effect of the outlet nozzle. The following values are known:

da = 0.064m di = 0.042 m

Thus dm = (0.064 + 0.042) m/2 = 0.053 m.

As already discussed in Chapter 2, the maximum peripheral speed of Ni-Cr steel (the basic material of the turbine wheel) is umax = 250 m/s, and the maximum blade temperature 600° C, provided that we are able to keep the temperature of the turbine disc substantially below that figure. As we will see, this is perfectly feasible.

The data we have already defined allows us to determine the maximum rotational speed nmax:

$$n_{\max} = \frac{u_{\max}}{d_a \cdot \pi} = \frac{250}{0,064 \cdot 3,14} \frac{1}{s}$$

nmax = 1243 m/s or approximately 75,000 rpm.

For further calculation we now need the turbine's outlet triangle and the mean peripheral speed u.

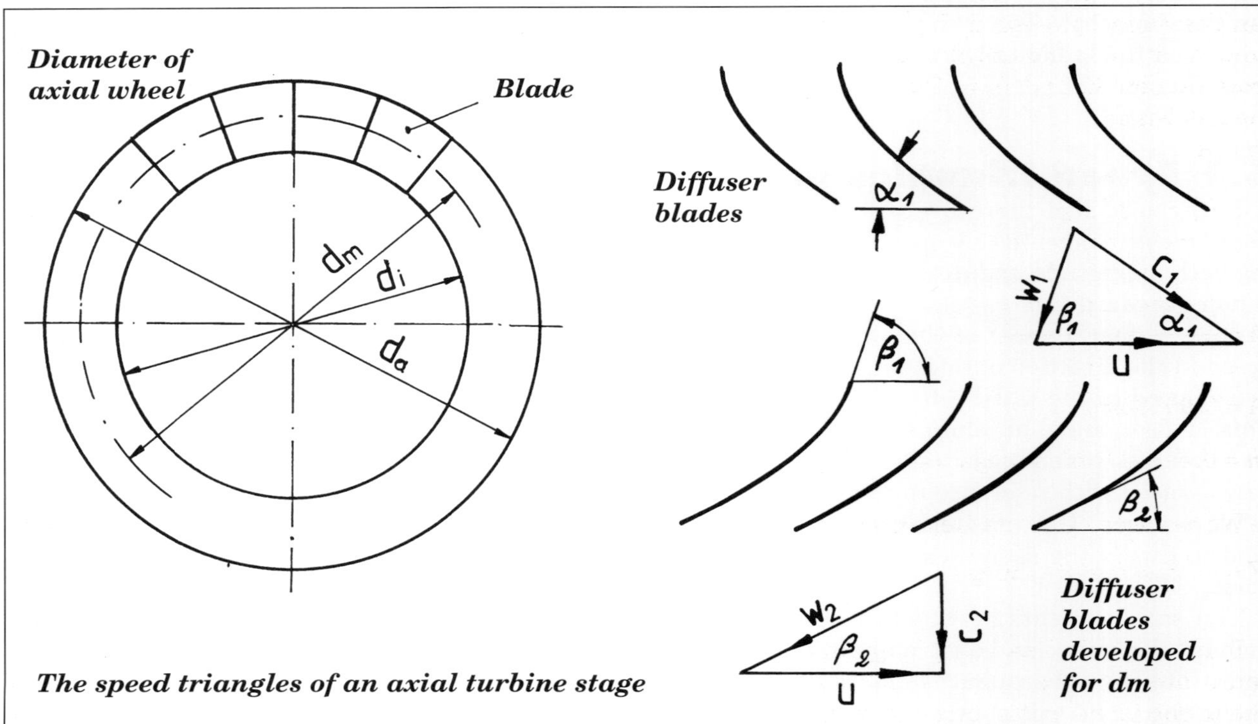

The speed triangles of an axial turbine stage

$$u = d_m \cdot \pi \cdot n_{max}$$

$$u = 0{,}053 \cdot 3{,}14 \cdot 1243 m/s = 207 m/s$$

At present we do not have a value for the outlet angle B2 of the turbine blades, which is 37°. This figure has been determined as a result of theory and practice, and is discussed in greater detail later. We can now draw the outlet triangle. We will assume a straight (i.e. not spiral) gas flow. This means: the outlet speed c2 is perpendicular to u, and therefore parallel to the rotational axis. c2 can also be calculated simply:

$$c_2 = 207 m/s \cdot \tan 37° = 156 m/s$$

Obviously this outlet speed is far higher than any practicable model speed. Therefore for our application it is not necessary to install an outlet nozzle with the intention of increasing speed. It is certainly possible to increase the thrust of the engine slightly, but more on this later.

As stated above, the gas temperature is 600° C = 873° K. We also need to know the mass throughput m, i.e. the quantity of gas in kg which passes through the engine. Since the exhaust gas is decompressed to ambient pressure, its density p is easy to calculate:

$$\rho_N$$

$$\rho = \rho_N \cdot \frac{273}{837} \frac{kg}{m^3} = 0{,}4 \frac{kg}{m^3}$$

Oxygen makes up only a small proportion of air, and in any case fuel combustion in the engine converts only part of the air's oxygen into CO2 and water, so the difference between the normal density of the exhaust gas and fresh air is insignificant, and can therefore be ignored. Similarly we can ignore the fuel mass flow, as it is much smaller than the airflow. From the outlet cross-section A, and from the outlet speed c2, we can calculate the following:

$$\dot{m} = A \cdot c_2 \cdot \rho$$

$$A = \frac{\left(d_a^2 - d_i^2\right) \cdot \pi}{4}$$

$$A = \frac{\left(0{,}064^2 - 0{,}042^2\right) \cdot 3{,}14}{4} m^2$$

$$A = 0{,}00183 m^2$$

$$\dot{m} = 0{,}4 \cdot 0{,}00183 \cdot 156 kg/s = 0{,}115 kg/s$$

We can now calculate the thrust F:

$$F = \dot{m} \cdot c_2 = 0{,}114 \cdot 156 kg \cdot m/s \cdot s = 17{,}8 \frac{kg \cdot m}{s^2} = 17{,}8 N$$

It must be obvious that, if d3 and dm are increased, and c2 and u are kept the same, thrust will increase at the same rate due to the larger outlet cross-section. If the peripheral speed u is viewed as a constant, then B2 also remains the same, but the rotational speed is reduced, as dictated by the formulae stated above. This is a good moment to remind you of this fact: stating rotational speeds is pointless unless we know the diameter of the rotating body!

Now we will pretend that we can vary the temperature of the exhaust gases and the blade angle B2 independently of each other. This will help us to understand the extent to which these values influence thrust, if all other conditions remain the same. We will therefore assume that the exhaust temperature is 100° lower. The density of the exhaust gas now alters by the factor 837 / 737 = 1.14.

The mass flow m, and also the thrust F, then increase by the same factor. The result is as follows:

F = 17.8 N . 1.14 = 20.3 N.

Lowering the exhaust gas temperature requires us to reduce the total losses in the system, or, in other words, to improve the overall efficiency of the engine.

Now let us increase the blade angle B2. We will assume an arbitrary increase to 40°, and leave everything else unchanged. Using the familiar equation, the outlet speed c2 then rises to 173.7 m/s. The mass flow also rises to 0.127 kg/s. Thrust can now be calculated using the other familiar formula:

F = 0.127 . 173.7 N = 22 N.

You can see that the blade angle is a very powerful factor in the thrust calculation, and can probably guess that it cannot be increased ad infinitum. As we have seen, as the blade angle B2 is increased, the outlet speed rises and thus the kinetic energy of the exhaust jet is increased. However, energy has to come from somewhere. In our system the only source is the combustion energy of the fuel. This entails higher exhaust gas temperatures, and a consequent rise in operating temperature of the turbine wheel as the angle B2 is increased. As a result we very quickly run up against the temperature limit of the material.

It is possible to describe the inter-relationship between energy conversion and flow in the turbine stage and the compressor stage with good accuracy. We will return to our speed triangle and explain first how the torque is produced at the turbine wheel. To simplify the calculation we will assume that the density of the gas does not change while it is flowing through the turbine. This is perfectly admissible since our turbine stage works at a low pressure ratio. To produce torque a force must be exerted in the peripheral direction. This force occurs as a result of the deflection of the gas flow through the turbine blades. The flow is ducted by the diffuser vanes so that it flows at an angle B1 and a relative speed w1. The deflection alone, in this case from the angle B1 to B2, exerts a force on the blades, in a similar way to a wing. This force is amplified by the acceleration of the flow from w1 to w2 in the blade ducts. While we are considering the power of the turbine wheel we only need to take into account the forces produced by speed changes in the peripheral direction. According to the physicist Euler the force is proportional to the mass flow multiplied by the magnitude of the deflection relative to the peripheral direction, shown as c1u. We can therefore calculate the turbine's power by the following formula:

$$P = \dot{m} \cdot c_{1u} \cdot u$$

We see that the speed c2, which is determined by the diffuser blades, plays a crucial part in defining the power of the turbine.

Long before model turbo-jets were developed, theoretical and practical investigations showed that flow losses due to friction and turbulation in all the blade ducts are at a minimum, when the energy absorption, i.e. the decline in pressure and temperature, occurs equally in the diffuser system and the turbine wheel. Thus in an ideal design the shape of the ducts between the diffuser and turbine blades would be the same, but with opposite camber. In this case B1 is 90°, and the speed triangles at the inlet and outlet of the turbine blades are identical but opposite (mirror symmetry). c1u then = u and the shaft power of the turbine is as follows:

$P_W = \dot{m} \cdot u^2$ In this case α1 also = β2.

The total power conversion in the turbine stage is much higher, of course. Another way of expressing this is to say that the input power PE is converted into useful power with an efficiency of nT. The following formula applies:

$$P_W = \eta_T \cdot P_E$$

The useful power consists of the shaft power, as just described, and the jet power, and we can express it in this way:

$$P_E = \frac{\dot{m} \cdot u^2 + \dot{m}\frac{c_2{}^2}{2}}{\eta_T} = \frac{\dot{m}\left(u^2 + \frac{c_2{}^2}{2}\right)}{\eta_T}$$

Using the speed triangle as already described, this formula applies:

$$c_2 = u \cdot \tan \cdot \beta_2$$

This expression can now be included in the above equation, and produces the following formula for input power:

$$P_E = \frac{\dot{m} \cdot u^2 \cdot \left(1 + \frac{\tan^2 \beta_2}{2}\right)}{\eta_T}$$

This formula shows that any increase in the angle B2 would very quickly lead to impossibly high input powers at the turbine stage.

We have not yet discussed the mathematical relationship between pressure and temperature and the input power of the turbine stage. This involves a two-stage calculation. We can calculate the compressor stage's pressure ratio from the known mass flow and the shaft power, although we must also take into account the efficiency of the compressor stage. The following formula applies:

$$P_W \cdot \eta_v = \dot{m} \cdot c_p \cdot T_1\left[\pi\left(\frac{\chi - 1}{\chi}\right) - 1\right]$$

In this formula the symbols mean:
nv = efficiency of the compressor stage
Pw = shaft power of the turbine wheel
m = air mass flow through the compressor and turbine stages
cp = specific heat of air at constant pressure. This value is

around 1000 J / kg K.

T1 = absolute temperature of inlet air in degrees K, e.g. 20° C external temperature (20 + 273) K.

pi = pressure ratio, that is the pressure in the compressor stage relative to the pressure of the air at the inlet.

X = ratio of specific heats cp / cv of air. This value is 1.4.

Since we can use Pw = m . u2 in the formula, m cancels itself out:

$$u^2 \cdot \eta_v = c_p \cdot T_1 \cdot \left[\pi\left(\frac{\chi - 1}{\chi}\right) - 1 \right]$$

This formula can be resolved according to pi.

The exponent $\left(\frac{\chi-1}{\chi}\right)$ has the value:

$$\frac{1,4-1}{1,4} = 0,285$$

$$\frac{\eta_v \cdot u^2}{c_p \cdot T_1} + 1 = \pi^{0,285}$$

$$\pi = \left(\frac{\eta_v \cdot u^2}{c_p \cdot T_1} + 1\right)^{\frac{1}{0,285}}$$

We can now calculate the pressure ratio in front of the turbine stage. We will use our original estimate of u = 207 m/s, ambient temperature of the air at the inlet T1 = 293 K and efficiency nv = 70%. This value for efficiency is based on my measurements of model compressors.

$$\pi = \left(\frac{0,7 \cdot 207^2}{1000 \cdot 293} + 1\right)^{\frac{1}{0,285}}$$

$$\pi = 1,408$$

For the moment we will not concern ourselves with the design of the compressor and how it actually works. All you have to do for now is to believe that it is about 70% efficient. If the external pressure in front of the turbine stage is 1 bar, then we will find an absolute pressure of 1.408 bar, or an increase in pressure of 0.408 bar.

To complete the description of the processes at the turbine stage all we lack is the temperature T3. This is the task: to calculate the temperature T3 for a given pressure ratio and known input power at the turbine stage. We use the following formula, which is resolved to find T3:

$$P_E = c_p \cdot T_3 \left(1 - \frac{1}{\pi^{0,285}}\right) \cdot \dot{m}$$

For Pe we use the known for-

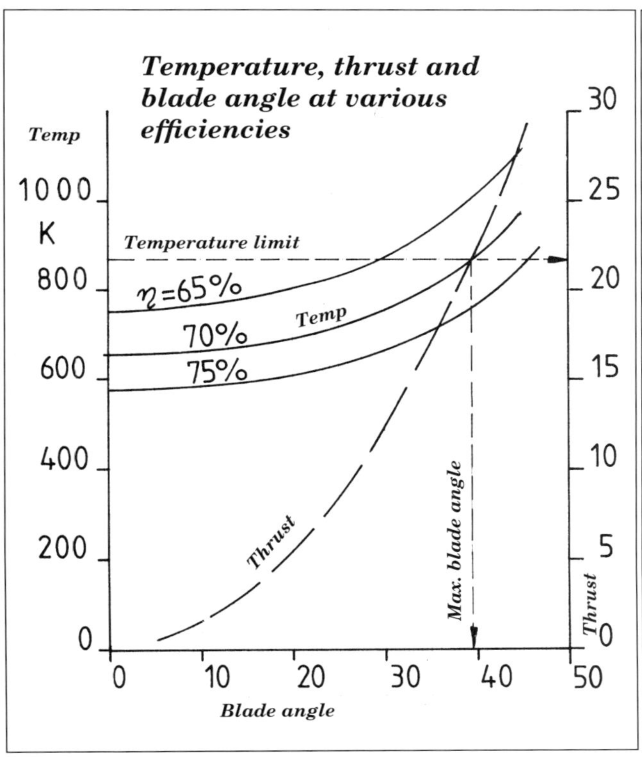

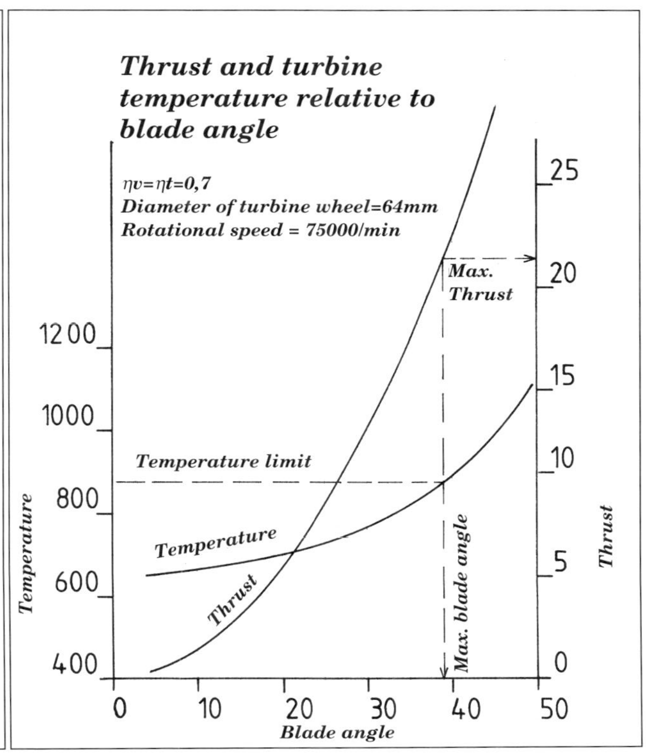

mula (1) and for pi the formula (2); we can then resolve it to find T3:

$$T_3 = \frac{u^2 \cdot \left(1 + \frac{\tan^2 \beta_2}{2}\right)}{\eta_T \cdot c_p \left(1 - \frac{1}{\frac{\eta_v \cdot u^2}{c_p \cdot T_1} + 1}\right)}$$

This formula is approximate, but is adequate for our purpose. It shows clearly that the temperature in front of the turbine stage must rise as the blade angle is increased. Note here that the derivation of the formula only applies to turbine stages with a reaction factor of 0.5, i.e. where the energy conversion occurs equally in the diffuser system and in the turbine blades. The graph printed here shows clearly the relationship between temperature and blade angle at different efficiencies. The maximum permissible mean peripheral speed is assumed to be 207 m/s, as already discussed. The inlet temperature is 15° C, or 288 K. Using our equipment it is not possible to measure the efficiency of the compressor and turbine stages with great accuracy. My own measurements show an efficiency of 0.7 for both stages, so the graphs have been drawn for efficiency values from 0.75 to 0.65 to illustrate the effects. For our purposes it is immaterial whether the compressor stage has an efficiency of 0.72 and the turbine stage only 0.68, or vice versa. Only if one stage is particularly inefficient does the formula fail.

The practical usefulness of this theoretical work has become very clear from my development of the "FD" series. I have built examples with blade angles of 30°, 37° and 43°, and the measured temperatures fitted extremely well on the 0.7 efficiency curve. If we accept 873 K or 600° C as the maximum permissible temperature, then a blade angle of 40° would still be feasible. The turbine blades of the 43 degree version became so hot at maximum speed that they were visibly distorted after a few minutes' running. The graphs indicate a blade temperature of 940 K = 667° C in this situation. With safety in mind, the blade angle of the FD 3/64 described in the building instructions has been kept to 37°. The measured and calculated graphs plotting thrust and rotational speed again confirm the usefulness of the theory, simplified though it may be. In terms of practical design the temperature variation relative to rotational speed is also of crucial importance. However, the temperature / rotational speed graph shows that temperature varies less in relation to rotational speed than thrust. In consequence the best method of increasing thrust for a given turbine wheel diameter appears to be to increase rotational speed. Unfortunately the stress on the material due to centrifugal force rises in proportion to thrust. Therefore an increase in rotational speed would require the use of exotic materials of high thermal strength for the turbine wheel. For the amateur modelmaker the technical problems which this involves are insoluble. As the graph of thrust / rotational speed shows, fitting a tapered nozzle (degree of taper found experimentally) can achieve a significant increase in thrust coupled with only a slight rise in temperature. In calculating the curves it was assumed that 10% of the pressure decline was removed.

If you want to build a more powerful turbo-jet but keep to simple tools and avoid technical problems, the easiest route is to make the turbine wheel larger. If you keep within the limits stated above for temperature, peripheral speed and blade angle, the maximum outlet speed remains the same. However, the cross-sectional area of the turbine rises by the square of its diameter, and air throughflow rises by the same amount. Thrust therefore rises at the same rate. Expressed simply, engine thrust rises by the square of the turbine diameter. On the other hand maintaining a constant peripheral speed means that maximum rotational speed is reduced in inverse proportion to the diameter of the turbine wheel. For example, if we make an 80 mm diameter turbine wheel instead of a 64 mm one, the maximum permissible rotational speed n80 falls:

$$n_{80} = n_{64} \cdot \frac{64}{80}$$

$$z.B. n_{80} = 75.000 \frac{64}{80} U/\min = 60.000 U/\min$$

On the other hand, thrust rises:

$$z.B.: F_{80} = 20N \cdot \left(\frac{80}{64}\right)^2 = 31{,}3N$$

If a nozzle is used it would probably be possible to extract 20% extra thrust, giving a final potential static thrust of 37 N. Naturally the mass of the engine would also be greater, since virtually all the components would have to be enlarged by the same degree as the turbine wheel. However, it is likely that you could expect slightly improved internal efficiency as a result of the enlargement.

At least in theory there is a possible method of increasing thrust for a given wheel diameter, temperature and peripheral speed, and that is by constructing a blade system which gives a greater deflection at the turbine wheel. Drawings A – D show the possible configurations. Note that it would take considerable extra effort to make turbine wheels with the blade forms shown as C and D. Nothing is known about efficiency at model sizes, and since no investigations of this kind have been carried out (to my knowledge), you would have to carry out your own experiments. In general terms we can say that any intermediate shape could be tried between the designs shown in the drawings A – D, and I would not expect a dramatic change in running characteristics with moderate variations on the calculated shapes. It goes without saying that for every configuration a purpose-designed compressor stage would be required. The extent to which inaccurate matching alters the engine's operating characteristics depends to a very great extent on the design of the compressor stage.

As a means of checking the design we need a method of calculating the exit speed c2 and the mass flow m. For this we need the measured static thrust and temperature values, as well as the cross-sectional area of the outlet behind the turbine. The density ρ of the exhaust gas can be calculated from its temperature. The following two formulae apply:

$$F = \dot{m} \cdot c_2$$

$$\dot{m} = A \cdot \rho \cdot c_2$$

$$c_2 = \frac{\dot{m}}{A \cdot \rho}$$

This expression for c2 is used in the first formula:

$$F = \dot{m} \cdot \frac{\dot{m}}{A \cdot \rho} \cdot \frac{\dot{m}}{A \cdot \rho}$$

Resolved to find m:

$$\dot{m} = \sqrt{F \cdot A \cdot \rho}$$

Equation 1 can now be resolved to find c.

As a practical example we will consider the FD 3/64. Measured values for thrust and temperature are F = 20 N and 630° C = 903 K.

$$\left(1N \mathrel{\hat{=}} 1\frac{kg \cdot m}{s^2}\right)$$

$$\rho = 1{,}29 \cdot \frac{273}{903} kg/m^3 = 0{,}39 kg/m^3$$

$$A = 0{,}00183 m^2$$

$$\dot{m} = \sqrt{0{,}00183 \cdot 0{,}39 \cdot 20 \cdot \dot{m}^2 \cdot \frac{kg}{m^3} \cdot \frac{kg \cdot m}{s^2}}$$

$$\dot{m} = 0{,}119 \frac{kg}{s}$$

Blade forms and speed triangles for axial turbine stages

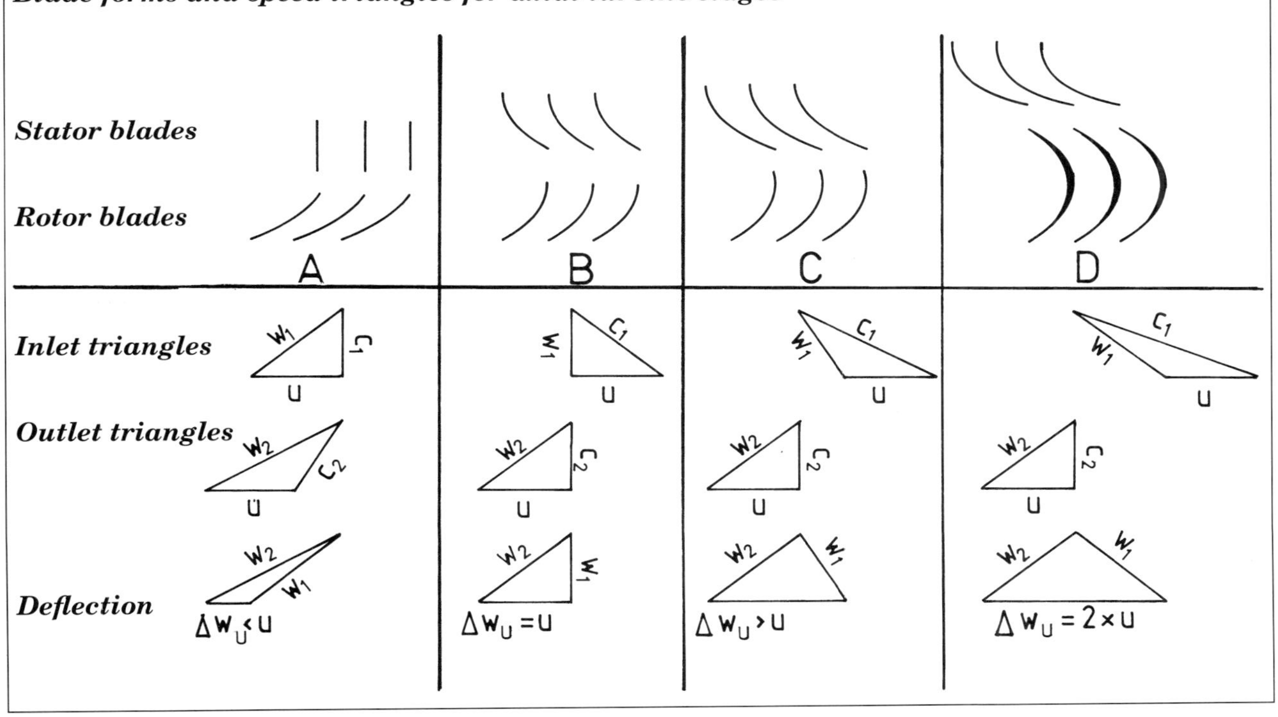

This value for m can now be used in the first formula to calculate c2:

$$c_2 = \frac{F}{\dot{m}} = \frac{20\dfrac{kg \cdot m}{s^2}}{0,119\dfrac{kg}{s}} = 168\frac{m}{s}$$

If we measure the rotational speed at the same time, then we have an extremely useful method of checking the calculated data. Using this method of calculation it is essential that the flow speed is approximately constant in the area under investigation. This method is equally suitable for calculating the outflow speed and the mass flow of ducted-fan engines. We can also measure the static jet power using the formula:

$$P_{Strahl} = \frac{\dot{m}}{2} \cdot c^2$$

This value in turn can be used to calculate the system's efficiency relative to the motor's shaft power. One final point: calculating the data for designing a turbo-jet, including the data relating to the engine's behaviour in flight (as described in Chapter 3) is much simpler than for a piston engine with a propeller or ducted fan.

The final step in designing the turbine stage is to estimate thermal expansion. The thermal expansion coefficient a of the materials used in the hot gas section of the engine are as follows: for steel 12 x 10-6 / K and for nickel-chrome steel 16 x 10-6 /K. To be completely accurate, these values vary with temperature, but for our purposes this is less important.

If we assume even heating, the length change can be calculated as follows:

$$\Delta l = l \cdot \Delta T \cdot \alpha$$

l is the length of the component under study, e.g. the diameter. ΔT is the temperature difference between the cold and hot states. In all normal metal alloys expansion occurs evenly in all spatial directions. In the turbine stage the only change which concerns us is to the diameter. The diameter of the FD 3/64's turbine wheel is 64 mm. We will assume the wheel's operating temperature is 600° C in the worst case. Thus ΔT = 580° C and the change in diameter is:

$$\Delta\varnothing = 64 \cdot 580 \cdot 0,000016mm$$

$$\Delta\varnothing = 0,59mm$$

This corresponds to a change in diameter of almost 1%, i.e. with a gap width ΔØ of around 0.3 mm the turbine blades still just have clearance if the wheel is centred accurately in the housing, and if we ignore expansion due to centrifugal force. In practice the centre of the wheel stays substantially cooler than 600° C, but on the other hand the turbine blades may get hotter during the starting process and under acceleration. The value as calculated above is therefore a realistic estimate for the maximum change in diameter. In fact the problem is made less severe by the thermal expansion of the housing. In the region of the turbine wheel the housing reaches a temperature of around 400° C, while the diameter is 66 mm. Using steel as constructional material this produces a change in diameter of:

$$\Delta\varnothing_2 = 66 \cdot 380 \cdot 0,000012mm$$

$$\Delta\varnothing_2 = 0,3mm$$

Under stable conditions, i.e. when engine temperatures have settled down, we can assume a mean temperature of 500° C for the turbine wheel. The corresponding change in diameter is then only 0.49 mm. Under stable conditions the outside diameter of the turbine increases by 0.3 mm more than the diameter of the housing. In mathematical terms this still leaves a gap width of 0.15 mm to spare. In practice a gap width of 0.4 mm, measured cold, is sufficient to avoid problems.

4.3 Design of the compressor stage

4.3.1 Design of the compressor wheel relating to airflow

In our calculations relating to the turbine stage we have established the maximum rotational speed, the mass flow m and the shaft power Pw, and can now attempt to design a suitable compressor. This is one area where specialist literature comes to our aid; for example, Willi Bohl's book entitled "Ventilatoren" ("Fans"). There is no distinct dividing line in technical or physical terms between fans and rotary compressors. In practical terms fans are generally seen as relatively low pressure flow machines, in contrast to compressors, which can be operated under conditions of high pressure.

The first step in the design process is to settle on the optimum type of compressor wheel for our application. First we calculate what is termed the running factor based on the operational data established during the design of the turbine stage. All fan experts are familiar with the Cordier graph, and from it the optimum wheel form can be read off directly provided that the basic data is known. The basic wheel type varies from radial flow via diagonal to axial flow. A typical radial wheel is the fan of a vacuum cleaner. At the other extreme of the Cordier scale is the axial fan, of which the conventional airscrew is a good example. In between we find the impeller fan, the multi-blade wheel of a multi-stage axial compressor and the fan of a large turbofan. There are no sharp distinctions between these different types. Bohl summarises the results of the Cordier graph, in terms of the optimum wheel type, in the following table:

Running factor	**Type of wheel**
0.06 – 0.8	Radial wheel
0.25 – 1.0	Diagonal wheel
0.6 – 3	Axial wheel
0.35 – 05	Drum rotor

The optimum type of wheel is the type which offers maximum efficiency under the stated conditions. As the table shows, the classifications overlap. For example, if the running factor is 0.5 it is possible to build a radial wheel and a diagonal wheel of approximately equal efficiency. The drum rotor in the last line of the table is of no interest to us because of its shape.

First we have to calculate the running factor using the formula:

$$\sigma = \frac{2 \cdot n \cdot \sqrt{\dot{V} \cdot \pi}}{(2 \cdot Y_t)\frac{3}{4}}$$

n = rotational speed in s-1

Yt = total specific fluid feed [lit. "fluid supply"] work in J/kg

For our application we can use the following expression from our turbine calculations to determine the specific fluid supply work.

$$\frac{P_w}{\dot{m}} = u^2$$

$\dot{V}$ = Volumetric flow through the compressor stage in m3/s. This can be calculated easily from the temperature and pressure on the compressor side and from the mass flow m. A quick reminder here: the mass flow is of equal magnitude through the compressor and through the turbine. In contrast, the volume flow varies according to temperature and pressure.

pi = 3.14

The calculated data derived from our estimates concerning the turbine stage were:

u = 207 m/s

m = 0.121 kg/s

n = 1250/s or 75,000 rpm

Using this data Yt = 42,800 J/kg.

We can assume 1.25 kg/m3 as the value for air density on the compressor side. Thus V = 0.097 m3/s. Now we can calculate σ as follows:

$$\sigma = \frac{2 \cdot 1250 \cdot \sqrt{0{,}097 \cdot 3{,}14}}{(2 \cdot 42800)\frac{3}{4}}$$

σ = 0.285.

This running factor clearly indicates the choice of a radial compressor wheel, even when we take into account some uncertainty in the individual values stated above. Thus we can state with confidence that our compressor can be built in one stage with a radial wheel.

However, we all know that virtually all full-size turbo-jet engines employ multi-stage axial compressors, so we need to discuss why this layout is unsuitable for our purposes. One reason is obvious: the large number of compressor stages involves greater constructional complexity. To achieve our very modest pressure ratio of around 1.4 an axial compressor would have to be of at least three-stage construction. My own measurements using a single-stage axial compressor of suitable size showed a maximum efficiency of 60%. Similar efficiency values have been measured with ducted fan engines, which are also good examples of axial wheels. It seems that it is impossible to achieve levels of efficiency comparable to those of full-size axial compressors at our small sizes and low Reynolds numbers. In contrast, the measured efficiency of a radial compressor stage was at least 70%. These findings confirm the notion that the axial compressor is probably only superior at quite large sizes. To the designer of full-size jet engines the axial compressor's smaller diameter and consequent lower drag is a crucial advantage, since these factors are particularly important at high flying speeds in the sonic and supersonic range. For model flying speeds, on the other hand, this hardly matters at all. I would not wish to state that it is impossible to construct a small working turbo-jet using a multi-stage axial compressor, but one thing is certain: it could not be made using amateur equipment and methods.

Unfolding and explaining the effectiveness of the different types of compressor wheel would fill several volumes, so we will limit ourselves here to the essentials of the radial compressor. The most important means of increasing pressure is to exploit the difference in centrifugal force as the air flows from the small-diameter inlet to the larger-diameter outlet. At the same time the radial wheel imparts to the air an outlet speed which is approximately the same as the wheel's peripheral speed. This high level of kinetic energy can then be converted into pressure energy. The conversion occurs in the stator of the compressor stage which works as a diffuser.

Let's stay with the radial wheel for the moment. Many variants of the radial wheel can be built: e.g. forward-curved blades, retro-curved blades, radial-tipped blades, with or without cover plate. In terms of efficiency the best design is a covered wheel whose blades curve backward. This type is the preferred choice for stationary fans. Wheels with radially-tipped blades produce the highest increase in pressure for the smallest diameter, but they are not so efficient. It is possible to compensate for this disadvantage by clever design of the diffuser system. This is the preferred design for professionally produced small gas turbines, e.g. APUs (Auxiliary Power Units). A number of professionally manufactured model turbo-jets have also been made using this type of compressor. My own calculations and experiments, bearing in mind the requirements of the engine in practice, have shown that the type with retro-curved blades appears to be the most suitable for our application.

The drawing below shows the main dimensions of such a wheel, together with the speed diagrams at the entry and exit edges of the blade, with the corresponding angular data. The main dimensions are important when we come to construct such a unit.

ds = inlet diameter; in practice this corresponds to the diameter d1 of the entry edge of the blades.
d2 = outside diameter
b1 = width of the blade channel at entry
b2 = width of the blade channel at exit
R = radius of blade camber
p = geometrical location of radius R
rs = radius of the cover plate at the inlet
s = thickness of blades.

The numerical data and all other information for making the

Principal dimensions and speeds of a radial compressor stage

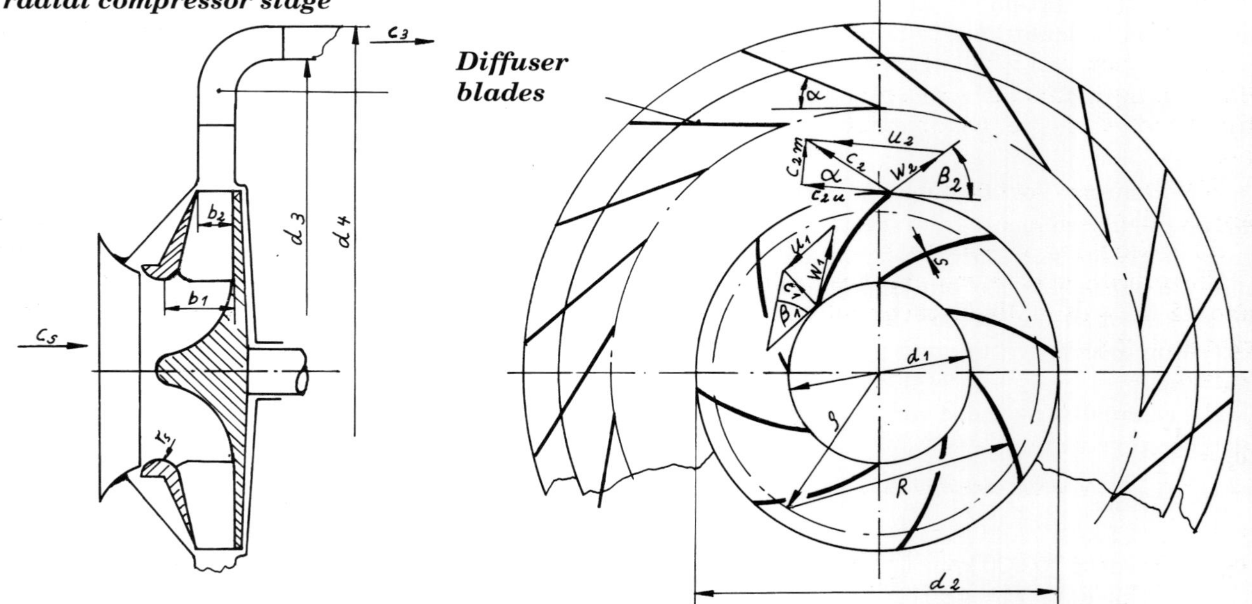

Shown without cover and cover plate

compressor wheel are included in the building instructions.

Since energy conversion is the primary purpose of the compressor, we must consider the speed and flow conditions in the whole of the compressor stage. Before we delve into the complexities of the speed plans, we must address the question: what specific work Yth can be converted in such a wheel with zero-loss throughflow? In other words: how much shaft power can such a wheel convert into pressure and kinetic energy in the airflow? According to Euler the following formula applies:

Yth = u2 . c2u

where:

u2 = peripheral speed at the outside diameter

c2u = projected tangential component of the exit speed c2.

It may appear surprising that that the inlet speed and the relative speeds w1 and w2 do not appear in this formula at all. The inference of this simple formula is as follows: for a given peripheral speed the work converted to arrive at the speed c2u can never be greater than Yth, regardless of how the air is supplied. As already explained, the converted work consists of the pressure energy and the kinetic energy of the airflow. Unfortunately the process of converting kinetic energy into pressure energy in the diffuser inevitably involves relatively high losses. Thus the essential art in designing a compressor stage lies in reducing those losses to a minimum. If a wheel with retro-curved blades is employed, the proportion of kinetic energy is smaller than for a wheel with radially tipped blades. This is why the first type is more efficient.

The less essential factors – throughflow, shaft power and rotational speed, and the derived maximum peripheral speed u2 of the compressor wheel – are already known. From this information it is simple to determine the maximum value for d2 from the formula:

$$d_{2\,\text{màx}} = \frac{u_2}{\pi \cdot n}$$

It is a moot point whether we need to push this value to the limit. In my opinion it is essential to bear in mind the strength of the wheel while we are attempting to optimise the flow conditions.

We now need to optimise the flow conditions. The variables are as follows:

d2, taking into account that d2 < d2max, b1 and b2.

The process of determining these values necessarily defines u1 and u2 as well as the throughput speed of the air cs, c1m and c2m and c2u.

$$d_1 \approx d_s$$

$$c_s = \frac{\dot{V} \cdot 4}{d_s^{\,2} \cdot \pi}$$ V = volumetric flow

$$c_{1m} = \frac{\dot{V}}{A_1} \qquad A_1 = b_1 \cdot \pi \cdot d_1 - \frac{s}{\sin\beta_1} \cdot z \cdot b_1$$

$$c_{2m} = \frac{\dot{V}}{A_2}$$

$$A_2 = b_2 \cdot \pi \cdot d_2 - \frac{s}{\sin\beta_2} \cdot z \cdot b_2$$

Where compression ratios are low, V can be considered as a constant at the input and output of the wheel.

$$c_{2u} = \frac{P_w}{u_2 \cdot \dot{m}} = \frac{Y_w}{u_2}$$

To the expert theoretician – those of you who understand fluid behaviour in detail – this formula may cause a groan or two. We already know that the inevitable losses mean that the full wave length can by no means be converted. Thus the correct expression for the denominator Yw would be Yw·η_v. We also have to bear in mind a further phenomenon. With a finite number of blades the airflow does not follow the blade camber accurately, and a circulatory flow is superimposed between the blades. As a result the direction of w2 is such that the angle B2 is effectively smaller than is geometrically the case. The specialist literature takes this into account by introducing the output reduction factor u. The correct formula would then look like this:

$$c_{2u} = \frac{Y_w \cdot \eta_v}{u_2 \cdot \mu}$$

If the compressor wheel is well-designed, nv and u are of approximately the same magnitude, and this is what justifies my initial simplification. The reduced output factor rises slightly as the number of blades is increased. Now we come to the optimum number of blades z. This is one matter where even the most experienced and practical theoreticians cannot decide. But don't worry – any number between 8 and 12 is good enough for our wheels. In my experience the number of blade is not very critical for our purposes.

The optimum values relating to wheel diameter and blade widths determined using formulae, tables and diagrams [1] should not be considered to be strictly mathematical solutions. These values may be optimum in terms of efficiency, but it makes obvious sense to deviate slightly from them in favour of greater strength, in order to withstand high rotational speeds. If we optimise the system logically in an effort to obtain maximum possible strength at high rotational speed, we arrive at uncovered wheels with radially tipped or only slightly retro-curved blades. If we use the formula stated above, then c2u = u2. This means that the kinetic energy of the airflow is considerably higher for the same theoretical quantity of work, compared with a wheel with retro-curved blades. We then need a complex design of diffuser system of relatively large diameter to convert the kinetic energy into pressure energy efficiently. A turbo-jet optimised in this sense would be shaped like a low, flat cake, with the wide side facing the direction of flow. That doesn't even look right.

I have carried out many experiments with modified wheels, and the result is the compressor wheel as described in the building instructions. The table below shows the data calculated according to [1] and the proportions of the wheel used in the building instructions, in relative and absolute terms.

Dimension	**Calculated according to [1]**	**FD 3/64 version**	**Relative dimension**
d_2	70	66	dmaxturb.
d_1	33	33	$0{,}5 \cdot d_2$
b_1	16	13	$0{,}2 \cdot d_2$
b_2	8,5	7,5	$0{,}11 \cdot d_2$
β_1	29°	34°	34°
β_2	39°	45°	45°
R	37,3	42,3	according to formula
ρ	24,2	30,1	according to formula
z	9	11	8 bis 12
s	1,1mm	0,88mm	$0{,}07 \cdot b_2$

$$R = \frac{d_2^2 - d_1^2}{4(d_2 \cdot \cos\beta_2 - d_1 \cdot \cos\beta_1)} \qquad \rho = \sqrt{R^2 + \frac{d_1^2}{4} - R \cdot d_1 \cdot \cos\beta_1}$$

The practical modeller will be wondering how critical the main data values are. To put it another way: what happens if the main dimensions vary significantly from the values stated in the table? To answer this question, here are a few examples from the development history of the "FD" series.

Case 1

The compressor wheel had 12 blades with an outlet angle of around 70°, b2 = 8 mm and d2 = diameter of the turbine wheel. Result: the turbo-jet tended to fluctuate in speed when started, and was very slow to accelerate from idle to working speed. The fluctuating effect is a result of flow break-away in the compressor and/or the compressor area. It manifests itself as a deep growling

Kurt's F100 on 'Finals'

sound, quite different from the usual sound of the gas stream. When the same wheel was installed in a different "FD" with a slightly smaller blade angle at the turbine diffuser system, the engine would not run. From this we can assume that if the compressor wheel is over-dimensioned, the system collapses.

Case 2

The compressor wheel had 8 blades, radially tipped, but with an outside diameter corresponding to 90% of the turbine wheel diameter. Result: the engine ran, but the turbine wheel glowed bright red. It was therefore impossible to run the engine at maximum rotational speed, as required for flight.

Case 3

The compressor wheel had 12 retro-curved blades with a 50° exit angle, but d2 = 85% of the turbine wheel diameter. This was the smallest wheel used in any of the experiments. Result: starting characteristics and acceleration to maximum speed was excellent. The exhaust temperature was below 500° C, but thrust at 75,000 rpm was only 10 N.

Case 4

The compressor wheel was constructed with a cover plate set plano-parallel to the base disc, i.e. the inlet and outlet width of the blades was the same. The number of blades was 12, the blade angle at the inlet 40° and the other values as standard. Result: using this compressor wheel the turbo-jet provided trouble-free flying, but thrust was reduced, and the temperature at the turbine stage was markedly higher.

To sum up: real problems only occur if the compressor wheel is over-dimensioned. This can result if the values for d2, b2 and also β2 are excessive.

4.3.2 Design of the diffuser system

The air exiting the compressor wheel in the plane of rotation has a high absolute speed of flow c2 even with retro-curved blades. Using our standard compressor wheel this speed is 170 m/s at full load. The kinetic energy of the airflow corresponds to about 30 – 40% of the converted shaft power. However, we should aim at a considerably lower speed at the point where the air flows through the combustion chamber. A reasonable target value for the inflow speed through the holes in the combustion chamber is about 50 m/s, and we should therefore aim to slow down the outflow speed c2 from 170 m/s to about 50 m/s, with the highest possible gain in pressure. The theoretically possible pressure gain ΔPLeit is:

$$\Delta P_{Leit} = \frac{\rho_m}{2}\left(c_2^{\,2} - c^2{}_3m\right)\cdot \eta_{Leit}$$

ηLeit is the efficiency of the diffuser system, and can be estimated at 70%.

ρm is the mean density of the air in the diffuser system. It is around 10% higher than that of the ambient air.

Thus the actual pressure increase amounts to about 12,000 pa, corresponding to 0.12 bar (1pa = 1 n/m2).

The deceleration of the airflow from c2 to c3 is achieved by increasing the cross-sections A2 to A3. A2 is the outlet cross-section of the compressor wheel.

$$A_2 = d_2 \cdot b_2 \cdot \pi$$

A3 is the annular cross-section according to the formula:

$$A_3 = \frac{\pi}{4}\left(d_4^{\,2} - d_3^{\,2}\right)$$

The following formula therefore applies:

$$\frac{A_3}{A_2} = \frac{c_2}{c_3}$$

The space between the outlet of the compressor wheel and the deflection area in the annular gap works as a plate diffuser. The component C2 is considerably higher than c3m. c3 includes all the spiral motion, and this is not reduced in the plate diffuser. It is the purpose of the diffuser blades to reduce the spiral movement. The inlet angle of these blades can be calculated from the following formula:

$$\tan\alpha = \frac{c_{2m}}{c_{2u}}$$

An additional correction of +2° has to be applied to allow for the constriction effect due to the thickness of the blades. In our diffuser system the correct angle for α is 24°.

The blade clearance in front of the entry to the plate diffuser has a compensatory effect on the airflow if the angle α is not quite correct. If we consider the behaviour of the airflow more closely, especially during the engine's acceleration phase, we find inevitable changes in the direction of flow at the outlet of the wheel, and also in the speed of flow. In practice a slight residual twisting motion in the airflow behind the blades, i.e. in the space between the diffuser and the combustion chamber, does not cause any problems.

I have not carried out systematic experiments to optimise the number of blades in the diffuser system, but one thing is certain: a diffuser system with our geometry cannot work without guide vanes, because of the spiral motion. As an alternative arrangement the guide vanes can be installed in the space between d3 and d4. In this case the inlet angle should also be approximately as stated above, and be curved in such a way that they end pointing in the axial direction. My first engines capable of flight (FD2 and FD 3/62) were constructed in this way. However, construction is slightly more difficult, and there are no practical advantages. In a case like this it makes obvious sense to opt for the technically simpler solution, as in the FD 3/64 as shown on the plans.

The helical housing usually employed as a diffuser system in stationary fans does not require guide vanes. However, the geometry of a helical housing makes the component too bulky for use in a model aircraft. My first experimental engine employed a compressor stage with a four-duct helical housing. This arrangement was used in the first model turbojet engine to run autonomously using the turbine wheel and compressor wheel as described here.

4.3.3 Strength of the compressor wheel

The high peripheral speed of this engine component forces us to consider more closely the loads upon it due to centrifugal force. A detailed description of this problem can be found in Bohl [1]. When we consider the compressor wheel we can see that the retrocurved blades are subject to a bending load, and we can calculate the bending force. If the blades are fixed on both sides, the following formula can be used to calculate the bending stress at the entry edge of the blade:

$$\sigma = \frac{\rho{:}u_1{}^2 \cdot b_1{}^2 \cdot \cos\beta_1}{d_1 \cdot s}$$

σ = bending load in N/m2
ρ= density of blade material in kg/m3
u1 = peripheral speed of blades at the inlet
b1 = width of blades
s = thickness of blades
d1 = diameter at the blade inlet

We can now calculate the bending stress at the entry point of the blades of the compressor wheel, as described in the building instructions:

d1 = 33 mm = 0.033 m
p = 1 mm
b1 = 13 mm
u1 = 0.033 . 3.14 . 1250 m/s = 129.6 m/s
B1 = 34°
σ = 700 kg/m3

$$\sigma = \frac{700 \cdot 129{,}6^2 \cdot 0{,}013^2 \cdot \cos 34°}{0{,}033 \cdot 0{,}001} \cdot \frac{kg \cdot m^3 \cdot m^2}{m^2 \cdot s^2} = 49{,}9 \cdot 10^6 P_a \triangleq 50N / mm2$$

The bending strength of various timbers with a density of 0.7 g/cm3 lies in the range 70 to 110 N/mm2. I have run a wooden wheel for brief periods at rotational speeds of more than 90,000 rpm, without the wheel suffering damage. Using a different wheel with a greater blade width b1 = 16 mm and d1 = 35 mm partial blade fractures occurred at the inlet edge at around 76,000 rpm, although the method of fixing the blades in the discs was not as thorough as on the wheel described in the building instructions. When making calculations we have to allow for some degree of uncertainty.

Now let us look at the bending load at the outer edge of the blades, i.e. at the wheel width b2. We can use the same formula again, but this time using the data applicable to b2:

$d_2 = 66mm, U_2 = 259{,}2, b_2 = 7{,}5mm, \beta_2 = 45°$

$$\sigma_2 = \frac{700 \cdot 259{,}2^2 \cdot 0{,}0075^2 \cdot \cos 45°}{0{,}066 \cdot 0{,}001} Pa = 28{,}4 \cdot 10^6 Pa \triangleq 28{,}4N / mm^2$$

As a result we can see clearly that the bending stress on the blades of our compressor wheel is greater at the inlet edge than at the outlet. Damage analysis in the many experiments I carried out whilst developing a sufficiently strong wheel confirms that this method of calculation gives a good starting point for our purposes.

Calculating all aspects of the compressor wheel's strength, especially of the joints between blade and disc, would be a much more complex matter. In any case the results would probably be less accurate than those obtained by experimentation.

However, it is possible to estimate the strength of the discs. For discs of equal thickness with a small central hole the maximum stress is σmax=0,83·ρ·u2.

Using wood of 700 kg/m3 density and a peripheral speed of around 260 m/s (the maximum permissible speed of our compressor wheel), the stress is around 39 MPa = 39 N/mm2. The maximum stress occurs at the edge of the central hole, and not at the edge of the disc as you might have thought. Wood can certainly cope with this theoretical level of stress.

However, in our case the discs are substantially weakened by the slots into which the compressor blades fit. If the blades and discs are butt-joined, i.e. the discs are unslotted, the wheel will still withstand a rotational speed of around 50,000 rpm, corresponding to a peripheral speed of 170 m/s. As might be expected, the glued joints between the blades and the unslotted discs proves to be the weak point. On the other hand, the maximum peripheral speed of this all-wood structure was not all that far from the engine's full rotational speed, so an obvious move was to attempt to reinforce the wooden construction. This is hardly an original idea – old wooden carriage wheels with iron "tyres" and wooden barrels with iron hoops got there first!

The tensile force which occurs in a rotating ring is easy to calculate using the following simple formula:

$$\sigma_{Ring} = \rho \cdot u^2$$

For example, if we assume the use of steel with a density of 7850 kg/m3 and our maximum peripheral speed of 260 m/s, then the tensile stress due to the mass of the ring alone amounts to 531 Mpa, or 530 N/mm2. Now we will repeat this calculation using carbon fibre as reinforcing material, i.e. a ring of carbon fibre is wrapped round the wooden wheel. The density of the carbon fibre is around 1500 kg/m2, and the tensile stress then works out at only 101 N/mm2. However, it is permissible to stress carbon fibre far in excess of 1000 N/mm2. It must be obvious that this material is the perfect choice as reinforcement for the compressor wheel. The result is the component as described in the building instructions.

Strength is not the only factor here, however. We also have to consider the elastic distortion due to centrifugal force. This is easy to calculate for a ring; the change in diameter is:

$$\Delta d = d \cdot \sigma_{Ring/}E$$

E is the modulus of elasticity, which can be considered as a pressure. The numerical values for steel are 2.1 . 1011 N/m2, and the same high value applies to carbon fibre. Based on the stresses we have already calculated using the formula stated above, these figures produce a change in diameter of 0.17 mm for the steel ring, but only 0.03 mm for the carbon fibre ring. Bearing in mind the high tensile strength of carbon fibre, it is easy to see that it offers a far superior reinforcing effect in terms of strength and form stability to any other reinforcing material. The decisive factor in the material's form stability under otherwise equal conditions is the material constant p/E. The smaller this value, the smaller the change in size for given dimensions and peripheral speeds. In this respect carbon fibre is superior to every other material. Aramid fibres (Kevlar), also considered to be high in strength, are completely unsuitable for reinforcing the compressor wheel. This material's advantages lie in items which require high strength coupled with high flexibility, e.g. components subject to shock loads.

Finally a brief note on strength reduction as a result of the temperature increase in the compressor stage. With our low compression ratio of 1.4 the temperature increase is around 30° C. The following formula gives the approximate figure:

$$T_2 = T_1 \cdot \pi^{0,285}$$

where π is the pressure ratio of the compressor stage. Problems certainly arise with the high compression ratios found in full-size engines, where compressor temperatures in the final stage may rise to around 400° C, but they do not apply to us.

4.4 Fuel consumption

4.4.1 Calculating the fuel consumption of the FD 3/64

Once we have determined the operational data relating to temperature and air throughput, it is easy to calculate the engine's minimum fuel consumption. We can assume that the major proportion of the thermal energy in a heat engine using a low pressure ratio is used to heat the working medium. The thermal output can be found from the specific heat of air, the throughflow and the temperature difference using this formula:

$$\dot{Q} = c_1 \cdot \dot{m} \cdot \Delta T$$

c1 = specific heat of air

ΔT is the temperature difference between the inlet temperature and the exhaust temperature. Using the data for our turbo-jet engine, we find:

$\dot{Q} = 0{,}115 \cdot 600 \cdot 1050W = 72.450W$
heat output!

When hydrocarbons such as diesel, petrol and similar fuels are burned, each gramme releases 40,000 J of heat energy. As a result we can expect a minimum fuel consumption per second of 72.450/40 g/s = 1.81 g/s. If the fuel has a specific density of 0.85 g/ml this corresponds to a fuel flow of

1.81 / 0.85 ml/s = 2.13 ml/s.

Using this method it is possible to estimate fairly accurately the minimum fuel consumption for the entire rotational speed range, based on the mass throughflow and the temperature of the exhaust gases. Measurements show slightly higher values than those calculated (see diagram, Chapter 5.7).

In calculating the fuel consumption the calorific value is the only decisive factor.

4.4.2 The operating parameters corresponding to optimum fuel consumption

If we assume that our model aircraft is flying at a speed of around 50 m/s – a reasonable figure – we can expect thrust to be virtually the same as the static thrust value. We now need to consider whether it is possible to optimise fuel consumption relative to static thrust. As our calculations have shown, the choice of blade angle can be varied within quite wide limits, and this factor influences the ratio of throughput to the shaft power of the turbine wheel. By this means it is possible to achieve a higher pressure ratio for a given throughput. As is widely acknowledged, the thermal efficiency of an internal combustion engine depends to a very great extent on the compression ratio and the internal efficiency of the energy converter. Unfortunately the basic requirement if we wish to improve energy conversion is the use of a turbine wheel with a greater degree of deflection, but this cannot be achieved using simple manufacturing technology.

Starting equipment for the diesel fuelled FD3 turbojets is simple. Either a modified hairdryer or a 12volt vacuum cleaner can be used to spin the motor to minimum rpm, and a small aerosol canister of propane/butane fuel to preheat the combustion chamber. The last item needed is a match or cigarette lighter, unless the internal ignition system is fitted.

Chapter 5

Measuring apparatus, measuring techniques and the analysis of measured results

A complete turbo-jet system is bound to include a certain minimum of measuring equipment. This section describes what to use and how to use it.

5.1 Measuring rotational speed

As already explained the maximum permissible rotational speed of our turbo-jet engine is 75,000 rpm. Chapter 9 describes why the maximum permissible speed of a turbo-jet is so important. Conventional rev-counters designed for use with piston engines cannot cope with this high speed range, but special tachometers reading up to 100,000 rpm are available through specialist dealers. These units are contact-free, feature a digital display, and resolve down to 1 rpm. You have to attach a reflective mark to the rotor in order to take a measurement. The simplest method of doing this is to paint the hub of the compressor wheel matt black, leaving just a narrow radial stripe unpainted. It is impossible to operate a turbo-jet successfully unless you can measure its rotational speed. It is not absolutely essential to check the rotational speed every time you start the engine, but a tachometer is indispensable when you are adjusting the fuel metering system or preparing the thrust / speed and pressure / speed graph.

5.2 Measuring pressure

The ability to measure pressure is useful in two areas of the system:

1. To check the pressure of the fuel pump. You will need a manometer with a measuring range of about 10 bar.

2. To measure the pressure at the compressor stage of the turbo-jet. In our design this pressure is a maximum of 0.5 bar. A manometer with a measuring range of 1 bar is just right for this. Measuring pressure at this point is a useful means of checking how the engine is functioning when you measure its rotational speed. As the thrust / pressure graph shows, there is a linear relationship between thrust and pressure, i.e. at half thrust or double thrust the pressure will fall or rise in the same proportion. All you need to do is measure the thrust and the pressure simultaneously at maximum rotational speed, and you can then consider this value for maximum pressure as a reliable alternative to actual speed measurement. The great advantage of this system is that you can check that the engine is running correctly when it is installed in a model, since it is quite easy to fit a manometer connection.

If you wish to continue with your own line of development, you can make a sensitive and accurate manometer at very low cost in the form of a U-tube manometer. The tube is mounted on a vertical board with a millimetre scale, and the shanks filled with water. This is ideal for measuring small pressure differences, e.g. for measuring the pressure loss between the compressor outlet and the inner combustion chamber.

5.3 Measuring thrust

The thrust of a turbo-jet is undoubtedly the figure which interests us most. A set of domestic kitchen scales with a spring movement makes an excellent thrust meter. This can simply be anchored down on its side, and the engine, fixed to a mobile carriage, allowed to press against the scales. Of course, the mobile carriage could also be your model aircraft, ready for take-off. The photo shows my test rig, with the pump and control systems mounted on the carriage for test purposes, whereas they would normally be installed in the model.

The scale of the thrust meter should read up to 3 or 5 kg. If you are keen on accuracy, please note that a 1 kg indication on the scales corresponds to a thrust of 9.81 N. A spring balance could certainly be used instead of the kitchen scales. All you need to do is connect the balance to the test rig or the model, just outside the exhaust stream.

5.4 Measuring temperature

Measuring temperature – the exhaust gas temperature in particular – is just as important as any other measurement when you wish to check whether the turbo-jet is working correctly. A thermal element made of Ni-CrNi is very highly recommended for this task, but iron-constantan thermal elements can be used as an alternative. You will also need a milli-voltmeter and a set of conversion tables. As an alternative, reasonably priced digital thermometers which can measure temperatures above 1000° C are available from specialist dealers. These units can be used to give very reliable readings of the temperature distribution and the temperature profile in the exhaust stream. Fluctuations at the periphery of the outlet surface of a magnitude of around 100° C are not that crucial, and may not affect the working of the system. More serious temperature fluctuations, known as hot-spots, are easy to detect as you can simply see them, provided that the turbine is not being operated in brilliant sunlight.

The safest method of viewing the turbine outlet with the engine running is to place a mirror slightly outside the exhaust stream at a distance of about 0.5 m. Angle the mirror so that you can clearly see the turbine outlet in the mirror from a position outside the danger area.

If you want to measure the mean exhaust temperature, the only reliable method is to funnel the exhaust stream into a sheet metal pipe, placed over the engine's outlet. You can safely assume that the exhaust gases are thoroughly inter-mixed at a distance of about 30 cm from the turbine, and that is the point at which the mean temperature should be measured. Note that measuring the exhaust temperature in this way does reduce thrust. This method of measurement is particularly useful when you are trying to optimise the

The "FD 3" on the test stand with all the auxiliary equipment. As in the design of the engine itself, the test stand is based on the use of generally available materials and tools. In the foreground: battery, starter fan, gas lighter, tachometer and thermometer. Kitchen scales are used to measure thrust. Behind it a standard gas cartridge with outlet valve as the source of supplementary gas for starting only.

All the components mounted on the carriage have been flight-tested. The arrangement is as shown in the illustration entitled "Overall arrangement of the turbo-jet propulsion system". In the foreground: pump, speed controller, pump battery, receiver battery and receiver and switch. The fueltank bears two marks used to measure fuel consumption. The smaller tank contains the lubricating oil for the bearings. The lighter spot in the turbo-jet's intake is a reflective marker for optical speed measurement, attached to the inside of the inlet lip.

exhaust nozzle. If the nozzle is too narrow, the exhaust gas temperature, and thus also the engine's temperature, will rise very quickly. As a general guide, do not let the exhaust temperature exceed 600° C. If the measured exhaust gas temperature is well below 600° C, then careful restriction of the nozzle will achieve a slight increase in thrust.

5.5 Measuring fuel consumption

Fuel consumption is a very important factor in the design of your model. For a given flight time turbo-jets consume fuel at a higher rate than piston engines. What we need to know is whether the volume or weight of fuel to be carried will substantially influence the model's weight. Let us take an average flight time of 5 to 10 minutes. Measuring the fuel consumption presents no technical problems: with the engine set precisely at a known setting, e.g. half-thrust, you simply measure the time it takes to empty a fueltank of known volume. In this way you can measure the fuel consumption as accurately as you wish. To obtain a more accurate idea of the engine's consumption it is a good idea to carry out these measurements at several different thrust settings.

5.6 Measuring the direction of flow at the jet outlet

Ideally the exhaust gas stream should leave the nozzle perfectly straight. If this is the case, the flow energy is being converted into thrust with maximum efficiency. To check this make a small wind vane from sheet metal and attach it to a length of wire. Hold the vane immediately adjacent to the annular jet outlet. It will deflect in the exact direction of the flow, and you can immediately see whether the direction of the vane is parallel to the axial direction of the turbine. Small variations do not have a marked affect on thrust. An angular error of 10° means a loss of thrust of only 1.5%.

5.8 Analysing measured results

In my development work on the model turbo-jet it was of crucial importance that I measure every parameter, and this section discusses a number of systematic measurements. First let us consider the thrust / rotational speed graph. The measured thrust and the calculated thrust varying with rotational speed are shown for engine configurations with and without the exhaust nozzle. As can be seen, the measured values are very close to the calculated curve. An analysis of the theoretical considerations and the actual course of the measured values shows that thrust rises in a close approximation to the square of the rotational speed. For example, if we take the calculated value of 4N from the graph at 30,000 rpm, and compare the value, say, at 60,000 rpm, then the thrust at the higher speed

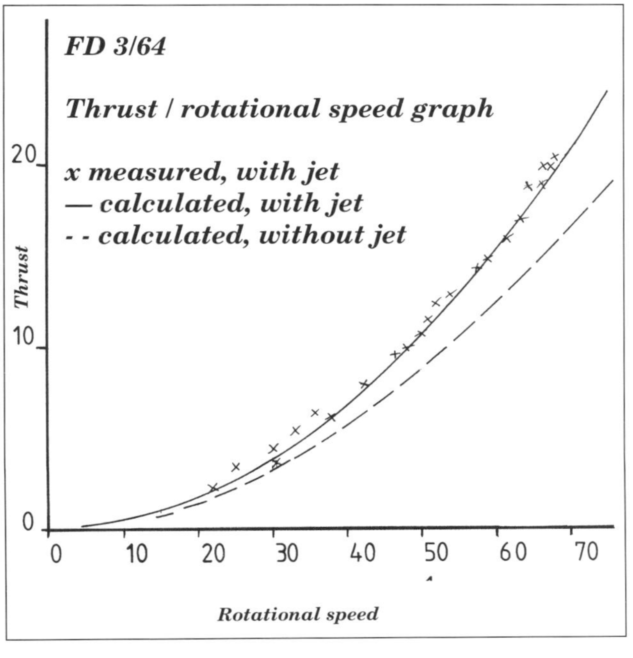

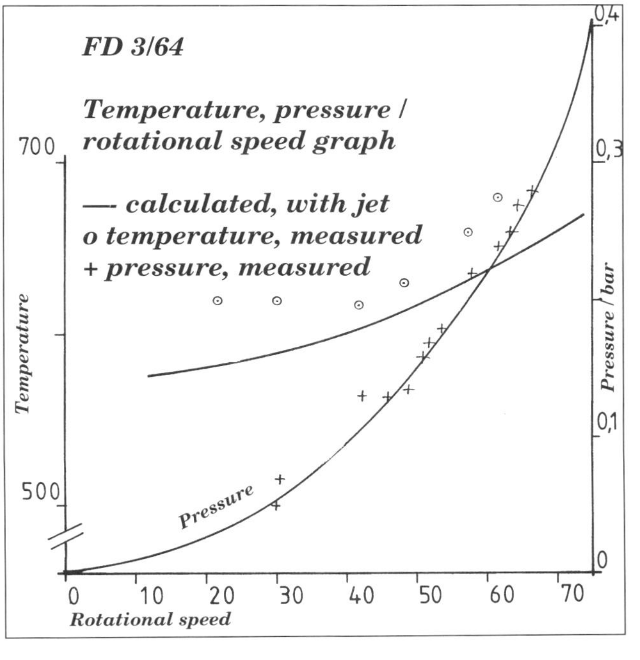

should be four times higher, i.e. 16N. The value taken from the calculated curve is 15.5N, which agrees very well with the measured values. That is an error of just on 3%. This result shows that our theoretical assumptions and methods of calculation coincide very well with what we find in practice.

Of course, the thrust / rotational speed graph is not sufficient on its own to prove the reliability of the theory for practical purposes. For example, it could be that although the calculated thrust / rotational speed graph agrees with the measured one, the engine's running temperature was very close to the critical value, or that fuel consumption was not even close to the previously calculated values.

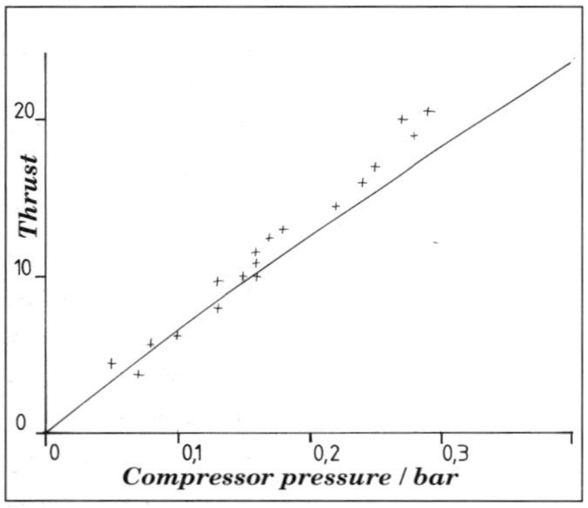

FD 3/64 – Thrust / pressure graph
x measured
__ calculated

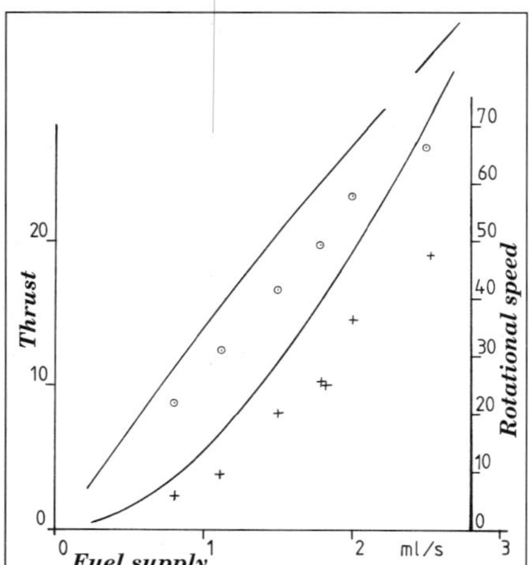

FD 3/64 – Speed and thrust relative to fuel supply
o Rotational speed, measured
+ Thrust, measured
__ calculated

For this reason we will now consider the graph for temperature / rotational speed, and compare the measured values with the previously calculated ones. We can see that the actual temperature in general terms is slightly higher than the theoretical curve. At low rotational speeds this discrepancy can be explained by the fact that the turbine simply cannot run at extremely low rotational speeds, i.e. it has a minimum running speed. Below this speed it would simply melt away at white heat if an attempt were made to increase speed. It could also indicate that the turbine's efficiency at low rotational speeds is worse than we expected. What is more interesting, however, is the temperature variation from the nominal curve in the upper speed range. The reason for this may be that combustion in the combustion chamber is less than complete. This has the following result: part of the fuel / air mixture burns outside the combustion chamber, and this in turn heats up the exhaust gas behind the combustion chamber. I will be able to say more on this matter when I have a chance to measure temperature distribution over the whole periphery of the annular nozzle. Major variations would indicate the fault stated above. In the series of measurements discussed here we did measure a temperature of 750° C at one point in the gas outlet, right at the annular jet. This could also be seen by observing the temperature distribution at the diffuser vanes. At dusk a very pale blue flame becomes visible at this point. This is a sure sign that a small part of the fuel is burning outside the combustion chamber.

If these inferences are correct, the results should be detectable when we compare the actual fuel consumption with the theoretically calculated consumption. As the diagram shows, this finding is quite clear.

Of particular interest is the course of the graph if we plot thrust against fuel consumption. As can be seen, thrust rises much faster than fuel consumption. The measured values confirm the calculated curve, but do lie slightly below it. However, rotational speed must rise in order to produce thrust, and that means: if we supply too much fuel, rotational speed quickly climbs into the critical area.

Finally we come to the thrust / pressure graph. In this case the measured values for thrust are slightly above the theoretically calculated ones. In practice these discrepancies are of no significance. If we turn back to the rotational speed / pressure graph, then we can easily see the maximum pressure which is permissible in order to achieve a thrust, say, of 20N, without having to use a tachometer. According to the calculated curve this would correspond to a pressure of 0.35 bar, measured at the compressor housing, with the engine running below the maximum rotational speed of 75,000 rpm. This keeps us on the safe side as far as rotational speed is concerned. The calculated thrust at this pressure is 21N, while the measured values indicate a slightly higher value.

If you are building a copy of the FD 3/64 you do not need not to measure everything as I have done, and as I have described. The point of this section is simply to illustrate how measured results can be analysed and put to use.

Chapter 6
Other Accessories

In addition to the range of measuring devices already described you will need a number of other essential items, although the exact inventory depends on your own preferences. If you only want to run your turbo-jet on the testbench, then you won't need a diesel fuel metering system and fueltank; instead you can use bottled propane or propane-butane gas as fuel. In this case the volume of the gas bottle should be at least 5 litres. Alternatively you can use small one-shot gas cartridges, but they are rather expensive, and not a permanent solution. The gas bottle or gas cartridge should be fitted with a fine regulator valve without a pressure reducer, and a high-pressure hose to connect the valve to the turbo-jet. Fuel hose intended for petrol makes a very good connecting hose; it should have an outside diameter of 5 mm and an inside diameter of 2.5 mm. This type of hose can also be used to connect the fuel metering pump when running the engine. If you intend running the engine on diesel fuel you will need at least a small gas cartridge fitted with a regulator valve as stated above. The "indispensable" category also includes a special oil tank and some form of starting equipment. A match or gas lighter is all you need to get ignition started.

6.1 Ignition systems

Combustion in the combustion chamber does not start by itself; it has to be set in progress once when the engine is started, after the auxiliary gas has been connected. Combustion then continues for as long as the starting gas or fuel flows. The simplest method of ignition is to apply a gas lighter or a match at the turbine outlet. Of course, it is possible to install a high-voltage sparkplug or a glowplug in the combustion chamber wall. Both systems work perfectly well. If you choose glowplug ignition you will need to pull its element out of the body slightly. Adjust the glowplug's power supply until the element glows yellow. The best position for the sparkplug or glowplug is the front part of the combustion chamber. However, you don't need to go to this effort if the turbine outlet is easily accessible.

6.2 Fuel metering system

The thrust of the jet turbine varies according to the amount of fuel supplied to it, so a suitable fuel metering system is required. A small electric-powered geared pump performs this task very well. If the rotational speed of the motor (and hence the pump) is varied, the fuel flow varies in the same proportion. The pump motor can be radio-controlled with the help of a standard speed controller, as used in RC cars and electric flight models. The arrangement is shown in the overall diagram. The carburettors of model diesel engines feature mixture adjustment screws, but this is not required for a turbo-jet.

No metering system specially developed for the requirements of the model turbo-jet is available commercially. However, it is not difficult to construct a reliable system using standard commercial components. You should bear the following points in mind when considering the fuel pump system:

1. Resistance to petrol, diesel fuel, petroleum and kerosene.
2. A leak-free drive shaft seal between pump and motor when under pressure.
3. Minimum pressure of 3 bar when pumping fuel at around 2 ml/s.
4. D.C. motor with an operating voltage of around 6 - 12 V with solid copper-graphite brushes.

I use a fuel pump intended for pit box installation, made by Conrad Electronic (Order No. 224421-22). The pump motor can be replaced by a more powerful unit, e.g. a Graupner RS 380. If you do this, remove the connecting piece from the original motor and press it onto the shaft of the RS 380. You will also need to fit a spacer washer (10 mm diameter, 1 mm thick) between the motor and the pump seal.

The Conrad catalogue also includes a useful fuel cock (Order No. 239321-22). This is a fundamental requirement, otherwise fuel will flow into the turbine when you are refuelling.

The speed controller should meet the following requirements:

1. Damped control characteristics, i.e. when the transmitter stick is moved quickly from zero to full load, the speed controller should only increase its setting gradually.
2. The controller's control characteristics should be as close to linear as possible with reference to the stick position. Some speed controllers increase speed gently, but have a very narrow control range between zero and full load. These types are completely unsuitable for accurate, constant fuel metering.
3. High pulse frequency (several kHz), otherwise the small pump motor may interact with the controller and cause interference to the radio control system.
4. The controller must not radiate interference frequencies (harmonics of the pulse frequency).

Most modern speed controllers designed for electric flight meet these requirements. The current consumption of the pump motor is only around 2.5 A, and as the operating voltage is also low, all current speed controllers will cope easily. EMF brakes and reverse running facilities are not necessary. I do strongly recommend that you carry out a range check of the radio control system with the speed controller connected to the motor, and set to all possible load settings. Note that even small electric motors must be fitted with suppressor capacitors.

A 5 - 6 cell NC battery pack with a capacity of around 500 mAh is adequate for the fuel pump. It makes no sense to power the pump motor from the receiver battery.

Another vital aspect of any reliable fuel system is a secure method of avoiding supplying too much fuel. The easiest way of achieving this is to install a restrictor in the feed pipe between pump and engine. The restrictor should be adjusted so that at the speed controller's maximum position, the fuel flow does not exceed the maximum permissible rate. This form of restrictor is easy to make, and consists simply of a piece of brass tubing of 1

Overall arrangement of the turbo-jet propulsion system

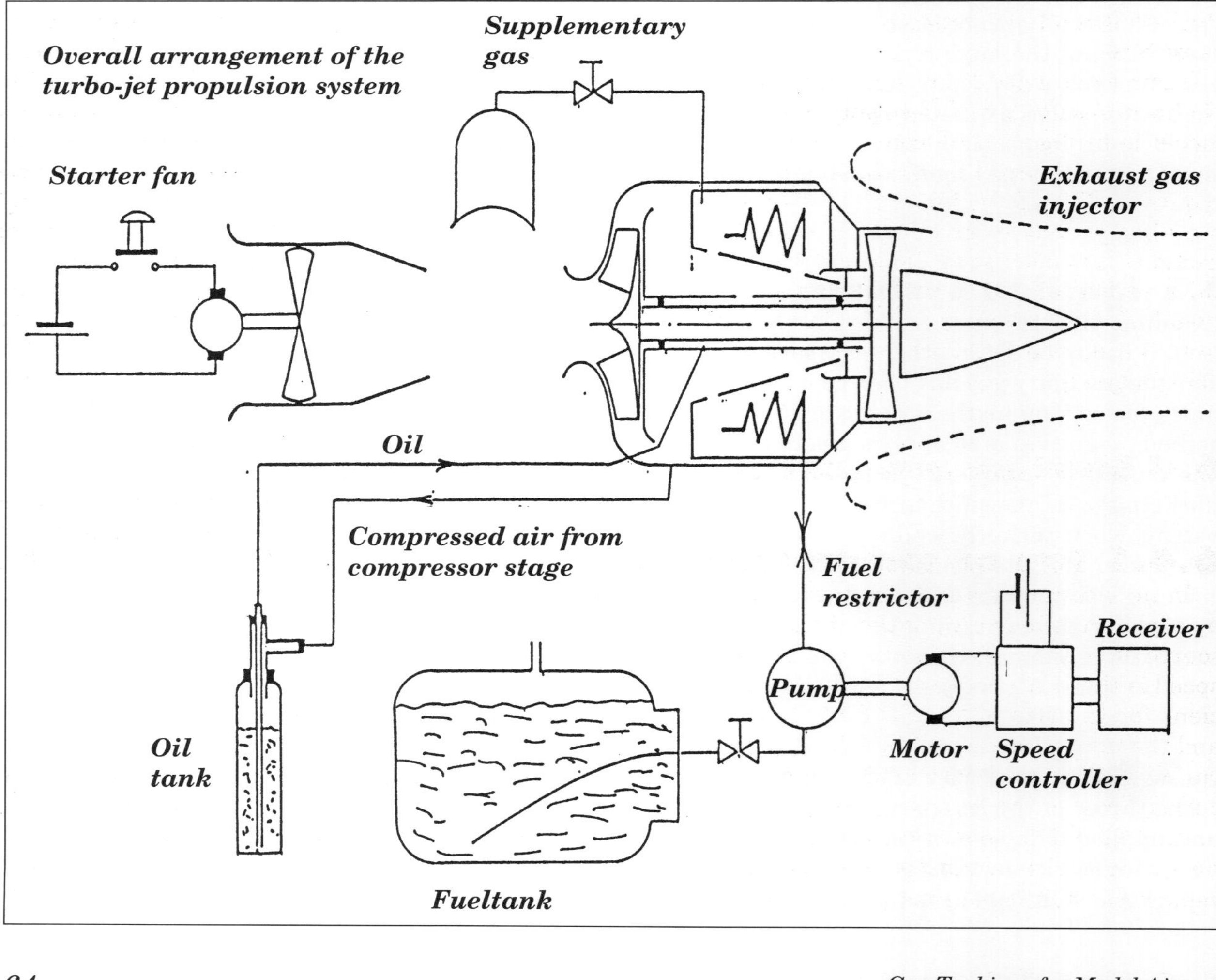

mm internal diameter and matching hose connectors. The tube should be around 10 cm long. A length of spring steel rod 0.8 mm in diameter is fitted inside the brass tube. Bend the rod to a gentle curve to prevent it slipping out. The length of this rod determines the effect of the restrictor; the correct length can only be found by experiment (see Operating Instructions).

6.3 Tanks

Two tanks are required: one for fuel, one for oil. Standard aerobatic model fueltanks with a capacity of 0.5 to 1 l are suitable. The only important point to check is that the flexible clunk hose should be made of petrol- and diesel-proof rubber. The connecting hoses to the pump must be of the same material. There is no need to pressurise the tank, as is often done with piston engines.

The oil tank does not need to be larger than 20 ml capacity. This tank should be made of transparent plastic, and must be installed in a position where the oil level can be seen clearly. The overall drawing shows the basic arrangement of the pipework to and from the tank. Since this tank is subjected to the pressure of the engine (max. 0.5 bar), it is essential to ensure that the sealing ring where the tubes pass into the tank cannot slip out. A small oilproof O-ring as seal retained with a wire loop similar to a champagne cork retainer have proved a good solution. If you cannot get hold of a suitable O-ring, you can cut a seal from thick-walled oil-resistant rubber tubing using a sharp knife or a razor blade.

There is no need to fit a clunk weight to the oil pick-up line. No harm results if the oil supply is interrupted briefly while the model is in flight. If oil consumption is excessive, use capillary tubing for the feed line to the engine, and adjust its length as required. The plastic inner tube of thin bowden cables (I.D. 0.8 mm) has proved an excellent material for this purpose. All connections between the oil tank and the engine can be made using thick-walled oil-resistant fuel tubing. You don't need an additional oil filler opening; simply disconnect the pipe between the tank outlet and the feed line and inject the oil into the tank using a syringe.

6.4 Starting equipment

6.4.1 Fan or compressed air

In an emergency a full-size turbo-jet engine can be re-started by exploiting the energy of the airflow alone, without any outside source of energy. The obvious way of spinning our turbine up to speed is therefore to use the airflow of a fan. A hair dryer is sufficient for the size of the FD 3/64. Remove the heating element and the protective cover first. The hair dryer's drive motor can be run directly from a battery of around 20 V DC voltage, and the current consumption is very low. Unfortunately you will need around 20 NC cells or two small 12-Volt lead/acid batteries connected in series to supply this voltage. Alternatively you can replace the hair dryer's motor with a model-type electric motor of the same size, e.g. a Graupner 380 RS. This motor provides sufficient power on only 7 - 8 NC cells, but it does draw a higher current. Replacing the motor saves the considerable expense of a high-voltage charger as well as a large number of NC cells. If you have some other type of electric fan to hand, you can quickly find out whether it is suitable for use as a starter: if the fan can spin up the rotor of the cold turbo-jet to a rotational speed of around 3000 rpm, it is good enough. Try fitting a nozzle onto the fan to see if this improves its efficiency as a starter. The distance between fan and jet turbine inlet is not critical. It can be a few centimetres, and even a few decimetres if the fan is powerful.

If the engine's inlet opening is concealed inside the model, the fan can be held directly to the fuselage air intake - this is usually sufficient to achieve the necessary 3000 rpm. If the model is fitted with two air intakes, one on either side of the fuselage or the wing root, it makes sense to provide an additional opening around 25 mm in diameter on one side of the fuselage in front of the engine opening. You can then fit a curved air duct on the fan and slide the duct through the opening for starting.

If you possess a powerful compressor, this can also be used as a starter. With a pressure of several bar the unducted flow of compressed air is enough to start the engine from a distance of half a metre.

6.4.2 Electric starters

An alternative method of spinning up the rotor to the requisite minimum speed is to use a small electric motor fitted with a flexible coupling. A short piece of thick-walled silicone hose makes a suitable coupling, but it must be a secure fit on the motor shaft. The coupling can then be pressed against the hub of the compressor wheel, just as if you were starting a piston engine. The power of the starter motor is not crucial. The decisive factor is its no-load speed, which must be substantially above the sustain speed of the turbo-jet. Small "buggy" motors, as used to power RC cars, are a very good choice here. The no-load speed should be at least 15,000 rpm. When you couple the motor to the compressor hub, without igniting the turbine, the rotor should run up to this speed, but not to the sustain speed. When the turbine is cold the motor's current consumption will be at its maximum. As soon as combustion starts, you will hear rotational speed rise clearly, and at the same time the motor's current consumption will fall. If you open the throttle further, rotational speed will continue to climb and the current consumption fall until the sustain speed is reached. At this point the motor ceases to supply power to the turbine, and motor current falls back to the no-load value. This sequence of events gives you a simple method of checking whether and to what extent the turbine is capable of running autonomously. If the motor is left coupled to the engine, the turbine will drive it above its no-load speed, and a current will flow in the opposite direction, i.e. the motor starts to work as a generator. You can switch off the starter motor once its current falls below the no-load value, and the turbine will then continue to accelerate under its own power.

You don't need to use this method to start a thoroughly tested turbo-jet such as the FD 3/64. However, if you follow your own line of development, it is the surest method of determining whether a turbo-jet is capable of running by itself. For example, if you ignite the engine but the starter motor's current drain remains the same or only drops slightly, even through the turbine's temperature is already high, you can assume that something is amiss with the system. The higher the starter motor's no-load speed above the engine's sustain speed, the more reliable the method.

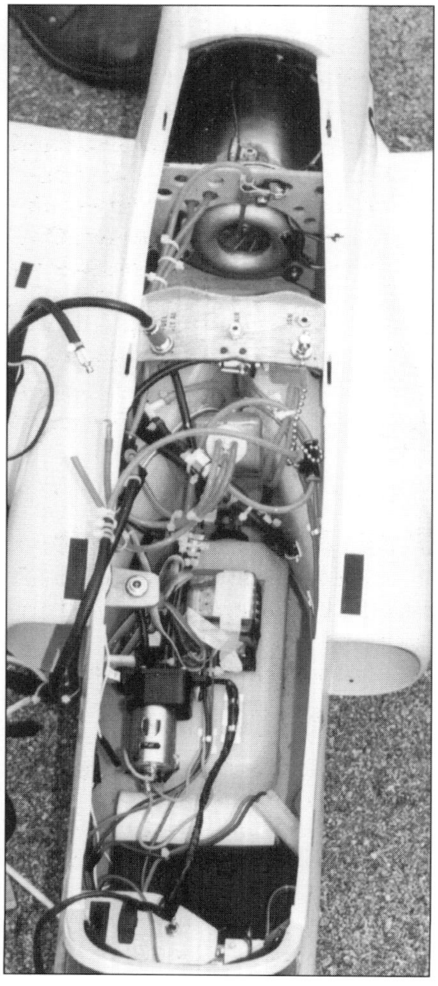

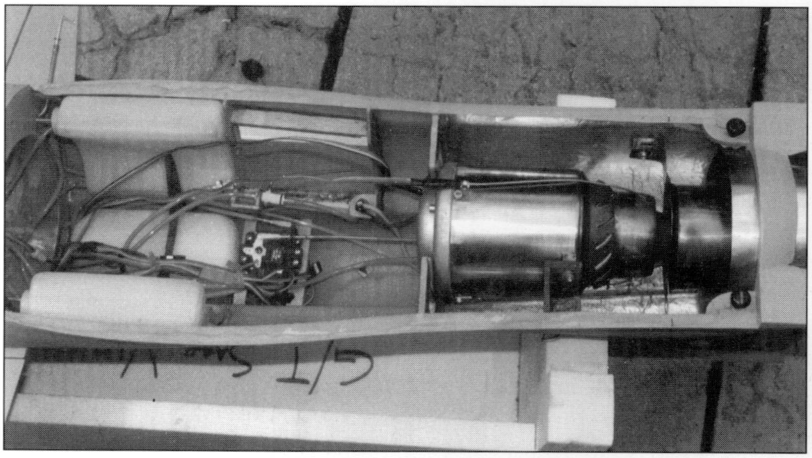

FD3/67LS installed in a Saab Viggen, showing location of the three separate fuel tanks and stainless-steel 'ejector' tube just behind motor.

Mike Koskela and Nick Moore shoe-horned an FD3/67LS and all ancilliary equipment into a BAE Hawk, as seen on front cover. Here you can see the fuel pump (front/left) and ECU on top of the forward fuel tank. The Hawk flies quite well, even at 14 lbs weight, on this turbojet - but landing and takeoff speed are rather high due to wing-loading.

Chapter 7

Building Instructions for the FD3/64 Jet Turbine

7.1 General information

I have personally made all the parts described in these instructions exactly as shown in the drawings, and all of them have been flight-tested. Before you start making your engine I recommend that you read right through these instructions in their entirety. The techniques employed have been selected with the aim of keeping technical complexity and difficulty of manufacture as low as possible. I have not attempted in any way to make the engine suitable for commercial manufacture. If you wish to make modifications to the design or construction, that is entirely up to you. However, please bear in mind that any change you make to the design, construction or choice of material does incur the risk that the whole system might fail to function.

Unfortunately it is impossible to make the rotor and some of the housing components without a lathe. The lathe must be equipped with the at least the following accessories: radially adjustable (three-jaw) chuck, tailstock with rotating (live) centre and a cross-slide with angular adjustment. Screw-cutting facilities for the common thread sizes are useful but not absolutely essential. The lathe must be large enough to accommodate the largest part to be turned or pressed. In this case this is the housing cover, whose blank diameter is 130 mm.

Certain parts must be made to a high level of accuracy, but I have designed them in such a way that expensive measuring tools such as a calliper gauge or internal limit gauge are not required. When the instructions state "front" or "front side", this refers to the air inlet side of the compressor wheel. Similarly "rear" and "rear side" refer to the gas exhaust side.

The complete rotor, ready to install

7.2 Constructing the components

7.2.1 The rotor system

This consists of the shaft assembly and the compressor and turbine wheels. The wheels cannot be made accurately unless the finished shaft is available. You will need the following measuring equipment: good vernier callipers, a dial gauge graduated to 1/100 mm and a screw micrometer. Vernier callipers with 1/100 mm digital readout can be used instead of a micrometer.

7.2.1.1 The shaft

This consists of the central shaft and the two bearing spigots. Turn down the central shaft to a diameter of about 14.5 mm, and leave it about 0.5 mm overlength. Bore the 10 mm holes and cut the M6 threads to take the bearing spigots. To avoid imbalance in the finished shaft I recommend that you check that the workpiece runs true using the dial gauge every time you have to re-chuck it. Adjust the centration of the chuck if necessary.

The permanent connections consist of M6 screwed joints and 10 mm nominal diameter bores, in which the bearing spigots are an interference fit. Measure the actual dimension and add 0.1 mm for the diameter of the spigot. Turn it down to this dimension, then chamfer the edge facing the central shaft. With the workpiece held in the chuck, attempt to screw the bearing spigot into the central shaft. It should be possible to screw the spigot into place as far as the 12 mm diameter flange using moderate force. If this is not possible, carefully grind down the diameter. Once you have successfully completed the screwed joint to the central shaft, pilot-drill the front face of the front bearing spigot (compressor wheel) using a 60º centre drill, and bore the 3.2 mm diameter hole for the M4 thread.

Turn down both ballrace journals 0.5 mm oversize. Do the same with the spigot for the M6 threaded section which accepts the turbine wheel. With both bearing spigots screwed to the central shaft, clamp the (uncompleted) turbine end of the workpiece in the chuck, and locate the other end on the live centre. It is extremely important that the following processes are completed without unchucking the shaft; this should ensure that the shaft, and thus the whole rotor, will run absolutely true. This is the sequence of operations: turn the tapered section at both ends, turn down the centre section to the nominal 14 mm diameter, then turn down the ballrace journals. The ballrace journals should be left 0.01 mm oversize. The ballraces should be a light force fit on the journals; this is achieved by polishing the bearing journals with the central shaft in the chuck. I recommend that you use the dial gauge to check that the bearing spigot is running true at the start of this process, and adjust the centring if necessary. Finish the job by cutting the M4 and M6 threads.

Note: it is also possible to machine the shaft from a single piece of steel. In this case the maximum diameter can be reduced from 14 to 12 mm without loss of bending strength.

7.2.1.2 The compressor wheel

This component is the most difficult part of the turbo-jet to make. The process is illustrated with photographs as well as drawings. In addition to the lathe you will need a small, precision pillar drill with a maximum speed of at least 6000 rpm, and a miniature hand-held grinder. If your lathe chuck cannot accommodate a blank of at least 50 mm diameter, you will need a 10 mm diameter mandrel and support flange to machine the inside of the cover plate.

The first step here is to make the bushes which are fitted to the base disc later. It is important that the bore (nominal diameter 8 mm) of the rear bush is a light force fit on the front 8 mm diameter bearing journal, to ensure that the complete wheel is accurately centred on the shaft. It is also important that the front face of the 10 mm diameter flange, which rests against the inner ring of the ballrace, is exactly perpendicular to the bore. The front bush is not machined to final shape until both bushes have been glued to the base disc.

Cut the blanks for the base disc and the cover plate from 6 mm aircraft-grade plywood, and turn them down to an outside diameter of 80 mm. Bore the centre 10 mm diameter. Cut an additional 50 mm diameter blank from the same material for the cover plate.

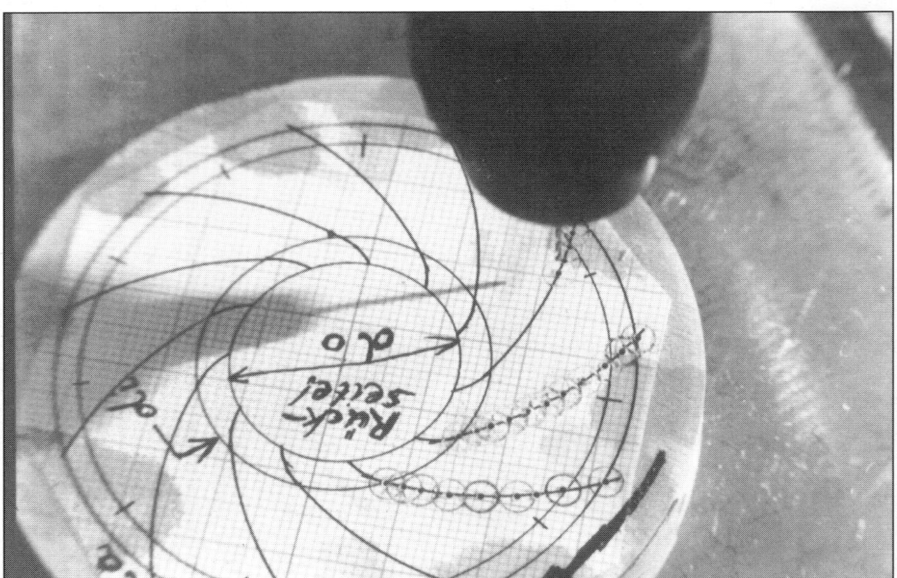

Marking out the blank and pilot-drilling the holes for the blade slots. The base disc and cover plate are drilled through in one operation between di and da. The base disc alone is then drilled between di and do.

Cutting the blade slots

Mark the centrelines of the curved blade slots on the rear face of the base disc. The easiest way to do this is to draw them on paper, then stick the drawing to the rear face of the base disc. Caution: the direction of rotation appears reversed when you look at the rear of the wheel. To allow for this you must draw the curvature on the template as a mirror-image. The angular division is 360º / 11 = 32.72º, but for our purposes it is sufficient to round the figure to the nearest 0.5º using a protractor, i.e. 32.5, 65.5, 98º ... etc.

Place the two 80 mm diameter blanks together concentrically and join them by gluing a few strips of wood across them around the outside diameter. Mark the front face of the cover plate and draw alignment marks across the edges with a pencil to establish the position of the discs relative to each other. The first step in cutting the slots is to drill a chain of 1 mm diameter holes spaced about 2 mm apart through both discs. The first holes should be drilled exactly at the inner diameter of 32 mm. Continue drilling until about 2 mm beyond the outside diameter of 66 mm. Separate the discs, then drill the 1 mm holes in the area from 26 to 32 mm in the base disc only. Connect the chains of holes using a fretsaw. On no account continue these cuts right out to the 80 mm outside diameter. Using a 1 mm bit in the pillar drill, machine out the slots using a high speed setting on the drill.

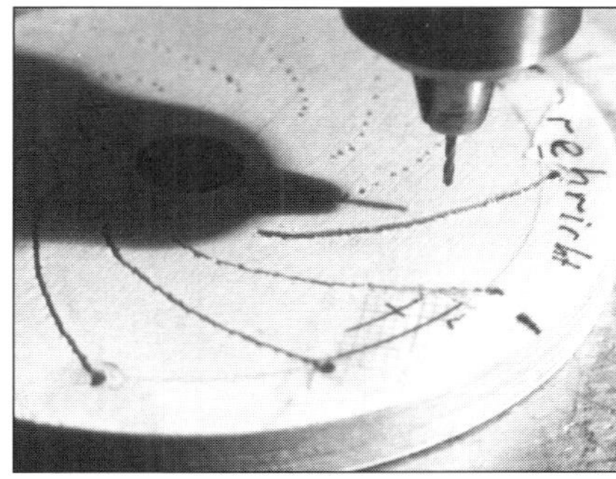

Milling out the slots.

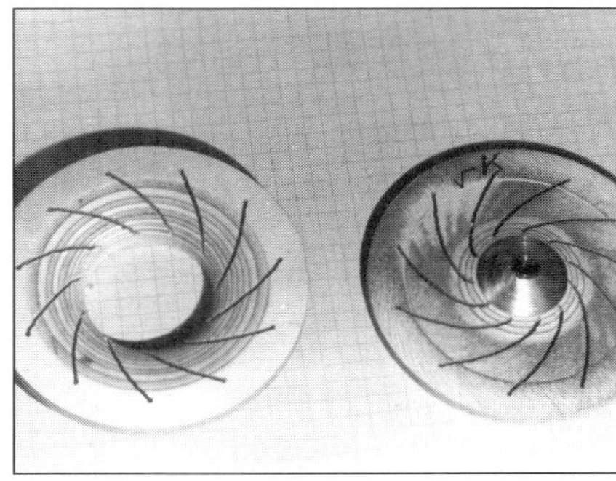

The machined inside surfaces after gluing the bushes into the base disc and joining the cover plate to the 50 mm diameter disc.

Installing the bushes in the base disc

The bushes can now be glued to the base disc. They must be a snug fit, and they should be fixed in place using heat-curable epoxy, e.g. UHU endfest 300. It should go without saying that the joint surfaces should be cleaned and keyed (roughened) before you apply the resin. Be careful not to apply excess glue. Fit a machine screw through both bushes and tighten a nut on the other side to hold the parts together under light pressure. Cure the epoxy at a temperature of 120º C, then let the assembly cool down. Fit the blank on the shaft with the ballrace in place, clamp it in the lathe chuck, centred on the ballrace journal, and tighten the retaining screw. You can now machine the bush to final shape, and machine out the stepped section to an internal diameter of 64 mm on the front face of the base disc. Machine out the curved face of the bush in small stages at first, then smooth it to a radius of about 17 mm using a mini-grinding disc. The actual radius is not crucial to the correct working of the compressor wheel. The only important point is that there should be a smooth, flowing transition from metal to wood, and that the worked surfaces should exhibit no wobble. It is essential to maintain the outer step as shown in the drawing, as it locates the carbon fibre reinforcement. It is sanded away when the compressor wheel is finished. Mark the position of the disc relative to the shaft, and always screw the parts together in the same position during the later stages of construction, and when you assemble the engine before running it. Leave the disc 80 mm in outside diameter for the present. Glue the 50 mm diameter blank concentrically to the front face of the cover plate blank using cyanoacrylate adhesive. The grain direction at the joint surfaces should cross at right-angles. Before applying the glue check on scrap material that it forms a strong joint with timber. When the joint has cured, clamp this assembly in the chuck at the 50 mm diameter. If your chuck cannot cope with this, screw it to the mandrel. Now you can machine out the inside shape of the cover plate as shown in the drawing. As

The three discs after gluing the blades in place.

The outer channels for the carbon fibre reinforcement are cut after the discs have been turned down to nominal diameter. The width of the inner flange of the channel on both discs is the edge "K"

Winding the carbon fibre reinforcement. The channel for the next smaller diameter should not be cut, and the carbon fibre wound into it, until after the first winding has cured.

with the base disc the 0.5 mm shoulder is important. Leave the outside diameter at 80 mm. Bore a central 33 mm diameter hole in the cover plate.

Preparing the blade blanks is simply a matter of cutting them to the size shown in the drawing. They should be left oversize in length and width. The stated material – beech 3-ply, nominal thickness 0.8 mm (= 0.88 mm) has proved an excellent choice for this purpose. Note that the grain direction of the outer plies must run parallel to the axis of rotation. Please don't try experimenting with different materials or thinner plywood – you have been warned!

Assembling the blades, the base disc and the cover plate

The important point here is that the two discs are plano-parallel and concentric to each other. Screw the base disc to the shaft with a ballrace in place, and clamp the shaft in the lathe chuck. Glue three small spacers exactly 6.5 mm thick to the internal edge of the cover plate, spaced apart by about 120º. Push three blade blanks into the slots in the cover plate which are adjacent to the spacers. Don't glue the blades to the plate yet. Push the other end of the blades into the corresponding slots in the base disc. Check that the alignment marks at the edge of the two discs are lined up correctly. The blades have to be curved to take up the camber of the slots. Since the blades are "sprung" in this shape, they stay firmly in place in their slots, with the result that, at this stage of construction, the whole assembly can be plugged together "dry" and dismantled at any time. Gently press the cover plate onto the assembly using the tailstock and a soft buffer, and check that both discs make contact with all three spacers. You are now in a position to adjust the concentricity of the cover plate relative to the base disc. Take your time, and be as accurate as you can. An eccentricity of 0.1 mm is satisfactory. Once you are confident that alignment is correct, glue the blades into the slots using cyano-acrylate adhesive. It is very important here to use a type of cyano which is expressly recommended for wood. Alternatively a slow-setting epoxy resin can be used. Once you have decided on the type of glue, keep to it when fitting the remaining blades. Once the joints have set hard, release the workpiece from the shaft and fit the remaining blades, working from the rear face of the base disc. Push them through until they rest against the 50 mm diameter disc on the front face of the cover plate. Glue them in place as described above.

Reinforcing the compressor wheel

Fix the compressor wheel to the shaft again, and turn it down to final outside diameter. The reinforcement consists of three annular windings of carbon fibre on each disc. Slow-setting epoxy resin is a suitable binding material, or you can use a specially formulated cyano-acrylate adhesive. In either case the important factor, as far as maximum strength is concerned, is that the carbon fibres are arranged very densely, are aligned accurately along the periphery of the wheel, and are saturated completely with the binding material. Fast-setting bonding materials, especially "instant" glues, do not meet this requirement. Glass fibres and aramid fibres are not a suitable alternative for carbon fibres. Aramid fibres are very strong, but at the same time they expand much more than carbon fibre, and their modulus of elasticity is much lower. The result would be that the reinforcing rings would expand considerably under centrifugal load, and the wooden component would fracture. Glass fibres also have a low modulus of elasticity and expand more than carbon fibres, and they are also more difficult to wind. We recommend that you practise the winding technique on a test disc of the same material and diameter. For the practice piece you don't need to cut the slots and machine the internal shape.

Now we come to applying the reinforcements: first turn the channels using a fine, sharp parting-off tool, then smoothe them using the finest grade of abrasive paper. The next step is to prepare the carbon fibres. The carbon fibre rovings generally available in model shops are too thick for our purpose, as the cross-sectional area is about 1 mm2. A roving of this type should be divided into about five strands, each of approximately the same thickness. If possible they should all be longer than 1 m. Splice the start and finish of each strand using a sharp balsa knife. This ensures that the individual fibres of each strand do not end at the same point. All strands must be free of knots and tangles.

Clamp the workpiece in the lathe chuck and wind the fibres onto the wheel by manually rotating the chuck, holding the carbon fibre strand taut, and guiding it by hand. Check that the start of the strand rests snugly in the bottom of the channel. You can achieve this by smoothing it down with a narrow strip of soft balsa. When the start of the strand is correctly positioned, apply a drop of adhesive. Apply glue successively at approximately each quarter-rotation. Smooth the end of the strand into the channel in the same way, and start winding the next strand at a point offset by around 180º from the end. Continue this process until the channel is completely filled with carbon fibres. Leave the bonding agent to cure for several hours before starting on the next stage, even if you are using cyano-acrylate adhesive. The next step is to cut the channels for the rings of the next smaller diameter.

Once you have completed the test piece, let the bonding agent cure then separate the carbon fibre ring from the wood on the lathe. You can now test the quality of the winding by carrying out a fracture test. Study the fracture point with a powerful magnifier, and you will clearly be able to see whether the fibres are fully saturated with binding agent as described above. If fairly large areas seem to be unsaturated, the adhesive or epoxy is not suitable. In my experience, UHU Endfest 300 and Simprop Blitzkleber "extra duenn" (ultra-low viscosity cyano) have proved outstandingly good. Although the Simprop adhesive is termed "Blitzkleber" (instant glue) it is sufficiently slow in response when applied to the carbon fibre windings using the method stated above, and permeates the fibres very well. I always recommend making an experimental winding in any case if you are using cyano-acrylate, since the adhesive characteristics of all cyanos vary markedly depending on their age. One of my compressor wheels constructed using this technique has survived rotational speeds of more than 90,000 rpm undamaged.

The finished compressor wheel.

A little trick is required to produce the channel for the smallest ring on the rear face of the base disc. Turn the step at the smallest diameter, then glue a thin plywood disc, 4 mm larger in diameter than the step, on the rear face. This produces the channel for the winding.

Finishing the compressor wheel

Part off the temporary plywood disc used for the final channel, then machine the outside of the cover plate and base disc to final shape. At this point all the surfaces, including those of the blades, should be impregnated with the adhesive you last used. When this has cured, sharpen the inlet edges of the blades using a fine grinding wheel in a mini hand-drill, and seal the worked surfaces with adhesive to complete the job. The transition areas where the blades meet the discs must be left unworked. Cut away the plywood at the outside diameter on the internal face of both discs, taking greater care to avoid damaging the blades and the carbon fibre windings. Ideally a thin layer of wood should remain between the carbon winding and inside face. Smoothe the worked edges using a fine emery board.

The final process on the lathe is to machine out and round off the inlet throat. The internal shape has to match the outside diameter of the inlet nozzle, and this cannot be done until the nozzle has been made. For this reason it makes sense to leave this stage until all the internal components and the cover have been completed. The actual process is completed in two stages: first bore out the inlet to the point where the inlet nozzle is a tight fit in it. Once the rotor and all the other parts have been installed, and the housing trimmed to fit, it is possible to centre the nozzle, the cover and the compressor wheel very accurately (see section 7.2.8). Once the parts have been centred, you can increase the diameter and depth of the opening by 0.2 mm. The worked surfaces must be impregnated with adhesive in the usual way.

Balancing the compressor wheel

This is best left until all the processes described above have been completed in full. Screw the compressor wheel to the shaft with a ballrace in place. Hold the outer ring of the ballrace between thumb and index finger to support the assembly, and run it up to a speed of about 10,000 rpm by playing the starter fan or compressed air on the compressor wheel. At quite a low rotational speed you will usually feel a distinct vibration through your fingertips – an indication of imbalance. Now apply a strip of fabric tape about 1 cm2 in size at any point on the outside of the cover plate, and repeat the test. Your fingertips will immediately tell you whether you have, by chance, found the right position, as the vibration will be noticeably weaker. Alter the position and size of the piece of tape and repeat the test until the vibration is at a minimum. This does not take long. Now remove the balance weight (the tape) and apply a drop of glue to the cover plate at that point, wiping it out to form a thin layer. Repeat the testing procedure as necessary. Your fingertips are quite accurate enough to balance the compressor wheel adequately. This method is so sensitive that your fingertips can easily detect a difference in mass of a few milligrammes at the edge of the compressor wheel, just by sensing the change in vibration. Once the balancing act is over, the compressor wheel is ready for use.

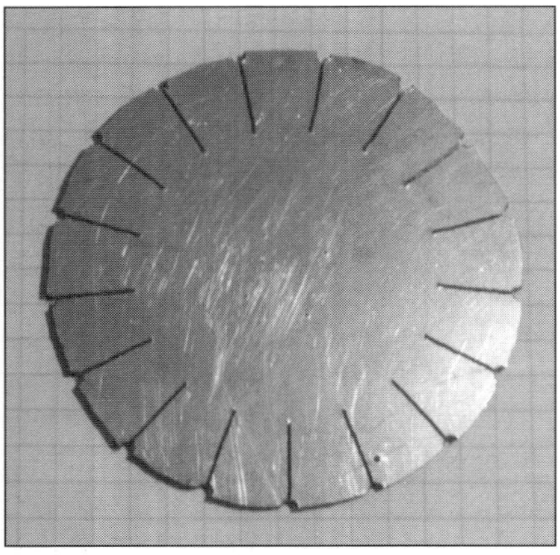

The blank after sawing out and marking the position of the flange.

7.2.1.3 The turbine wheel

To construct this component you will need MIG (Metal Inert Gas) welding apparatus and a small, robust, high-speed handheld grinder in addition to the lathe. Mark out the blade division on drafting paper using a root diameter of 41.5 mm and an outside diameter of 65 mm, and glue the template to the blank (2.5 mm thick stainless steel sheet). Mark the diameter of the mounting flange on the template. The stated outside diameter is about 1.5 mm larger than the diameter of the finished turbine wheel.

Mark both ends of the blade division slots on the metal at the inside and outside diameters using a centre punch. Mark the points on the outside diameter deeply, but those at the root diameter very delicately. Punch four points lightly around the hub flange diameter, spaced out at 90º to each other. Remove the paper template and mark the blade divisions from each punched point at the root diameter out to the corresponding point on the outside diameter, using a scriber. Don't scribe lines right across the disc, as this produces a serrated effect.

Making the blank

Drill 2.5 mm diameter holes at the outside point of each division. You will need to use a HSS drill in conjunction with a lubricant designed for stainless steel. In an emergency lubricating oil forms an adequate lubricant. Remove rough edges from the drilled holes. You can now cut out the prepared blank from the sheet material. Using a good hacksaw this takes about 30 minutes. Guide the saw blade from the centre of one hole to the centre of its neighbour, leaving half of each hole visible on the blank. Cut along the dividing lines between the blades, again using the fretsaw. Start from the half-holes and cut along the scribed line to the punched point at the root diameter.

Twisting the blades

Twisting the blades with the help of a claw tool.

Flange screwed to shaft and centred up with the blank prior to welding.

The next process is much easier if you first anneal the blank at about 700º C. Allow the workpiece to cool down fully. Clamp the blank in a vice between two 41 mm diameter steel rings. The front face of the blank, bearing the hub diameter marks, should be facing you. Locate the blade which is pointing vertically upward, and grip it in the claw tool as shown in the photograph. Twist the tool through 37º. Please be sure to twist the blade in the correct direction: looking at the blade from above, the twist should be anti-clockwise. To check the angle of twist use a protractor or an adjustable square set to 37º. Rest one shank of the adjustable square against the outside edge of the blade. If the angle is correct, the other shank will lie parallel to the vice jaws. Rotate the blank between the two rings until the next blade points vertically up, then repeat the procedure.

When twisting the blades please note that they should be twisted around their vertical axis only. They must not be bent in the direction of the axis of rotation. This can easily happen if you use pliers instead of the claw tool shown. It is a simple matter to twist the blades with an accuracy of better than +/- 1% using this method.

Making the hub

Clamp the hub blank in the lathe chuck and bore right through for an M6 thread. Bore the 6.5 mm mating hole, and turn the other dimensions leaving the workpiece 0.5 mm oversize. Cut the M6 thread and part off the hub. The corresponding shaft spigot can now be turned down to achieve the correct fit in the hub. The parts should be a light force-fit. Screw the two parts together, clamp the shaft in the lathe chuck, and turn the hub down to final size. Don't forget to check the centring of the ballrace journal beforehand, using the dial gauge.

Welding the hub to the turbine blank

Leave the shaft and hub chucked in the lathe. Press the turbine blank against the flange using the tailstock live centre and a cylindrical spacer, and check its alignment with the outside diameter of the hub flange. The four punched points on the blank are your reference points. The blank can now be attached to the flange by eight spot-welds. Use Cr-Ni wire of 0.6 mm diameter as a welding rod, using the MIG welder, of course. The spot welds should be exactly opposed to each other, i.e. offset by 180º. Cover the lathe under the joint area with metal sheet to avoid blobs of hot welding rod burning the machine. Turn down the blank to within 0.5 mm of its final outside diameter.

Grinding the turbine blades

For this process you will need a high-speed miniature hand-held grinder, protective goggles and a dust mask. If you are new to this process, start by practising on a

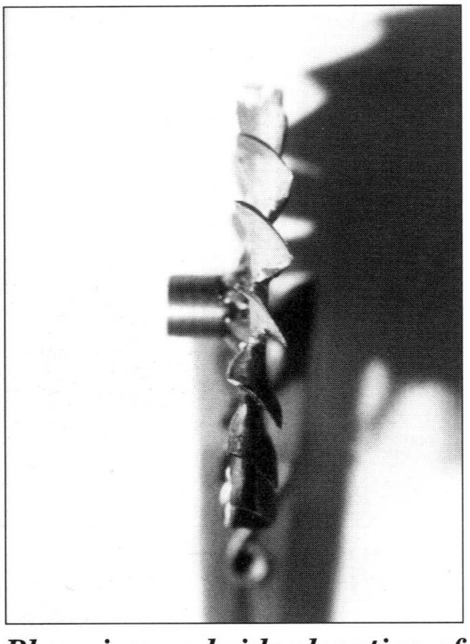

Plan view and side elevation of the finished turbine wheel.

piece of spare sheet metal – the same material used for the turbine blank. The grinder should be set to a speed where plenty of sparks fly when grinding.

The blades should be ground down to a shape which corresponds approximately to the cross-sections shown in the drawing. Exact fidelity to the profile shown is not absolutely essential. The profiles drawn in the cross-sections are designed with adequate blade stiffness and strength as the main priority. With this in mind, it is important that the blade root cross-section should be no thinner than that shown.

The first step in the grinding process is to open up the gap at the blade root to form a channel about 2 mm wide at an angle of 40º, measured relative to the axis of rotation. At the same time grind away any traces of the sawcut at the blade root. The diameter on which the base of these channels lies is 41.5 mm, as shown in the drawing. Grind the rounded area of the rear face of the blade to the radii shown in the drawing. The next step is to profile the front face. Note that the profile thickness reduces constantly from the blade root outward. The profile thickness at the outer part of the blade should be about 1 mm. Grind the front edge of the blades to a point, but leave the rear edge about 0.2 mm thick. Rounding off the front edge is not advantageous in terms of airflow. Finally grind down the tip of the blades to a width of 10.5 mm, finish off the taper at the front edge and, if necessary, adjust the blade thickness at the rear edge.

Balancing the turbine wheel

Before balancing the turbine wheel, turn it down to its final diameter of 63.5 mm. To do this screw it to the shaft again and clamp the shaft in the lathe chuck. Balance the wheel using the method described for the compressor wheel. However, in this case any imbalance can only be corrected by grinding material from the turbine wheel. It is usually the case that one blade or other has been left too thick, or too little material has been ground away at the blade root. It is a very quick matter to achieve an accurately balanced turbine wheel in this way, using the fingertip method. Do not under any circumstances attempt to balance the wheel by drilling into the disc or by grinding material away from the turbine disc itself. A turbine wheel with a hole in it, even if the hole is only part-way through, is completely unsuitable for high rotational speeds, and will certainly fail in use.

7.2.2 Jigs

Jigs are required to centre the housing component 14 relative to parts 15 – 18, and to the internal structure, parts 7 – 13. The first jig is a flanged disc similar to the turbine wheel, with the same diameter as the internal diameter of part 18 (jig A). You will also need two discs with the same diameter as the internal diameter of the housing, and a central bore matching the shaft diameter. Link these two discs together using three studs, spaced about 70 mm apart. You can turn the discs from 4 – 5 mm thick plywood. The two large discs with the studs form jig B. Jig C is used to align parts 15, 16 and 18.

7.2.3 The internal structure

The internal structure consists of parts 7 to 13. It's purpose is to locate and centre the rotor bearings in the housing, and it simultaneously forms the diffusor system for the compressor wheel at the front end.

We will begin with the shaft sleeve, part 7, which includes the bearing sockets at either end. The rear bearing socket is an easy sliding fit. The front bearing socket forms a fixed bearing, and provides the axial location of the rotor, in conjunction with the base plate, part 10. The bearing centres up parts 7 and 10 relative to each other.

The easiest method of making the bearing sockets is to turn them from bar stock initially, fit them in the tubular shaft sleeve, and hard-solder the parts together. Turn the shaft sleeve to length, then attach the front mounting flange to it using about 6 spot welds. The struts between flange and shaft sleeve can now be welded in place. Clamp the workpiece in the lathe chuck, true up the front face of the flange perfectly flat, and turn the front bearing sleeve as shown in the drawing. The position of the vent hole on the periphery of the shaft sleeve is not crucial, but it must be the correct distance from the front face, and the inlet point for the oil pipe, part 35, should be offset to the hole by 180º. Hard-solder part 35 to part 7, and attach it to one of the struts by wrapping with wire. Hard solder the joint.

Cut out the base plate, part 10, oversize. Don't use thinner material to save weight – that would be a false economy! The base plate forms a very rigid assembly in conjunction with the shaft sleeve, the housing and the link pieces, part 11. This rigidity is a fundamental necessity if the engine is to function correctly.

First drill through the centre bore in the plate and turn the ballrace seating. If your lathe does not allow you to chuck the base plate directly, screw it to a simple flat-faced flange for machining. The screw holes can be sealed later with countersunk aluminium rivets.

Make the link pieces, part 11, leaving them overlength, so that they can be cut down to suit the exact diameter of the housing. The threaded holes for the retaining screws, part 8, should not be drilled until the internal structure has been completed, and is ready for fitting in the housing. Rivet the link pieces to the base plate, using countersunk rivets on the front face of the base plate.

Saw out the compressor diffusor blade holder, part 12, to approximate size, and slot it to take the diffusor blades, part 13. Use the template method to mark out the slots, as described in chapter 7.2.1. Glue parts 10 and 12 together at this stage using heat-curable epoxy, e.g. UHU Endfest 300, and cure the joint at 120º C. Offer up this assembly to the assembly consisting of parts 7, 8 and 9, using a ballrace to aid location. The screw holes can now be drilled in the following sequence:

Drill the 2.5 mm through holes in the positions shown. Drill out these holes in part 8 to 3.1 mm diameter. Tap an M3 thread in parts 10 and 12. Harden the threaded holes in part 12 by applying cyano-acrylate glue. This produces a self-locking thread which holds quite well. For the next stage the two sub-assemblies need to be screwed together. Clamp the shaft sleeve in the lathe chuck, and centre up the other end on the ballrace using the live centre. This ballrace should only be used for the following machining process. Don't use it when running the rotor. Turn the diameter of parts 10 and 12 to final size. Machine the radiused edge and turn the front face of part 12 to size.

The next step is to install the diffusor blades, parts 13, in the slots, but you will need to remove excess adhesive from the slots beforehand. Use a miniature 1 mm diameter mill for this, or a 1 mm drill bit and a miniature hand-held drill. Position the threaded holes (for the bolts, part 23) so that their centres intersect the centreline of the adjacent blades.

Cut the diffusor blades, parts 13, to approximate size, sharpen their front edge (the edge which meets the airflow), and glue them to part 12. Use UHU Endfest 300 again, cured at 100º C. It is important not to use a higher curing

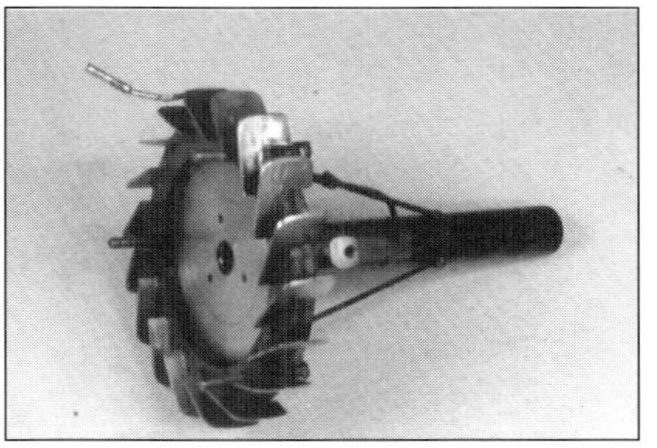

The completed internal structure. The tube projecting at top left is the oil pipe. The fuel and supplementary gas pipes are routed through the same area between the diffusor vanes. Three 3 mm diameter holes are provided at the corresponding points in the front housing. This pipe arrangement saves a lot of work with pipe sockets

The rotor installed in the internal structure. In this state it is possible to check the quality of the bearings before final assembly.

temperature this time, otherwise the diffusor blade holder may become detached from the base plate.

Parts 11 and 13 have to be finished on the lathe, but this can only be done if the housing, part 14, and the cover, part 25, have already been prepared. The first step is to turn the outside diameter of parts 11 and 13 to match the internal diameter of the housing, part 14. Round off the machined outside edges of parts 11 and 13 with a file, to make insertion easier. The parts should be a close sliding fit in the internal diameter of part 14, so that the internal structure can be pushed into the housing without requiring great force. Machine the 67 mm internal diameter of part 12, and chamfer the edges. Impregnate all the worked surfaces of the wooden component with fuel-proof lacquer. Machine the outside shape of the diffusor blades 13 to match the shape of the cover. Machining the blades is easier if you set up the lathe to rotate in the direction opposite to normal, and clamp the turning tool with the cutting edge at the bottom.

7.2.4 The housing

The housing is made from an empty Camping Gas International gas cartridge, type CV 470. The first step is to mark the 79 mm diameter on the valve end and cut along the line using a miniature cutting disc. Cut off the bottom section leaving the housing 108 mm long, measured from the rear opening. The best tool is again a miniature disc cutter. Sand or carefully burn off all the paint on the inside and outside of the cartridge. The painted finish on the inside and outside surfaces is now replaced with aluminium spray paint, available from many car accessory shops. When the

On the left, the housing blank. In its final stage it is transformed into a component fitted with integral diffusor vanes, turbine housing, mounting brackets and oil pipe (bottom right).

sprayed coating has dried, burn it on using a gas torch. At this point the mounting brackets, parts 36 and 37, the connecting nipple, part 42, and the reinforcements, part 39, can be hard-soldered to the housing. The best solder to use for these joints is low melting point silver solder.

7.2.5 Turbine diffusor blade system and turbine housing

This assembly consists of the diffusor blade holder, part 15, the diffusor blades, part 16, the central body, part 17 and the turbine housing, part 18. The first step is to make part 17 as shown in the drawing. You can save on work by making the outer cylindrical ring from tubing, cut the inner disc from 0.8 mm thick steel sheet, and weld it to the outer ring. Apply aluminium spray to part 17 and heat-treat it when it is dry. Alternatively this part can be welded up from nickel-chrome steel sheet, in which case no surface treatment is required. The radius at the front edge is non-critical.

Cut parts 15 and 18 to size from 0.8 mm thick sheet steel as shown in the illustration, and bend them to form a cylinder and truncated cone respectively. A piece of wood turned to the inside shape makes a useful former for shaping these parts. When the shaping stage is completed, weld parts 15 and 18 together using the MIG welder. This is the procedure: hold the parts together on the jig, tack-weld the parts at a few points along the seam, then remove the jig and weld the seam completely on the outside. Clean up the inside of the joint between the two parts using a miniature hand-held grinder. Apply the usual surface treatment to this component: aluminium spray fixed by burning on.

Cut the diffusor blade slots in part 15 working from the outside, using a miniature disc cutter and hand-held drill. Mark the slot positions using the paper template method. Glue the paper template in place securely and cut the slots using a grinder or drill, cutting along the marked centrelines. Make the wooden jig C as shown in the sketch before installing the blades. It is important that the 64 mm diameter of the jig is a press-fit in part 18. The jig forms a stop for the inside edge of the diffusor blades, part 16. The diameter of the stop should be approximately 1 mm smaller than the outside diameter of part 17 at this point. The result is a pre-defined excess at the inside edge of the diffusor blades, ready for final working.

The first step in making the diffusor blades is to cut a pattern from 0.5 mm sheet brass. This blade should fit as accurately as possible against the cylindrical part of the jig, and project by about 2 mm beyond the outside edge of part 15. The curvature of the outside edge of the blade is dictated by the shape of the slots in part 15. The radius of the curve should become steadily smaller towards the cylindrical part of the jig, although exact adherence to the stated radii is not a pre-requisite for the correct functioning of the system. Once this template is completed, you can make the actual diffusor blades using it as a pattern. The blades are installed as follows: fit the blade through the slot and tack in place on the outside using the welder, then adjust it if necessary. The blades should be radially symmetrical when viewed from the front and rear, but aligning them by eye is quite accurate enough. When you are satisfied, weld the blades to part 15 along the length of the joints, using the MIG welder. Use 0.6 mm diameter steel wire as welding rod.

Press out the jig, and clamp the diameter of part 18 in the lathe. The inside length of the diffusor blades can now be ground down to final size using the miniature grinder. Part 17 should fit inside the diffusor blades with about 0.2 mm clearance.

7.2.6 Attaching the diffusor blade system to the housing

Push jig B into the housing. Screw jig A to the shaft instead of the turbine. Place the sub-assembly consisting of parts 15 to 18 on the rear face of part 14, and slide the shaft in from the rear using jig A to centre up the parts. The sub-assembly can now be hard-soldered to part 14. Alternatively, if you are confident of your ability to weld very thin sheet metal you can weld the joint instead. Remove the jigs and clean up the hard soldered or welded seam on the inside of the joint. Use brass-based hard solder for this joint.

7.2.7 Centring the internal structure

Fit the front bearing in the internal structure by pushing it into the housing from the front. Fit a ballrace on the turbine end of the shaft, then screw jig A in place again. Slide this assembly into the shaft sleeve from the rear. This process accurately centres the shaft sleeve relative to the turbine housing. Measure the dimensions carefully, and drill the retaining screw holes through part 14 and the link pieces, part 11. Drill 3.2 mm pilot holes for the M4 thread first, then remove the internal structure from the housing. Drill out the holes in the housing to 4 mm diameter, and tap an M4 thread in the holes in parts 11.

7.2.8 Making the front section

This consists of the cover, part 25, the reinforcing ring, part 26, the connecting ring, part 28, and the inlet nozzle, part 27. Parts 25 and 27 are pressed out of 1 mm thick pure aluminium sheet. This technique is very easy to learn. I suggest that you start with the simpler component: part 27. You will need a hardwood former turned to shape; alternatively you could assemble the former from a stack of plywood discs. The outside shape of the former should be the same shape as the inside of the component to be pressed, but should be slightly longer, as shown in the drawing. Clamp the former in the lathe chuck. The workpiece blank is now pressed over the flat front face of the former. The live centre is used for this, in conjunction with a pressing disc whose diameter is slightly smaller than the front face of the former. The blank is a flat disc of sheet metal with the diameter stated in the drawing. It must be annealed at around 300º C before being shaped. Instead of a turning tool, a pressure tool is used for the next step. For our purposes this consists of a length of hardwood about 10 x 10 mm in cross-section, with the front face rounded off. For this simple type of pressing a lubricant is needed, e.g. grease. The principle of the process is this: the former rotates, and the sheet metal, in its soft state, is pressed against the former using the pressure tool. The first step is to guide the pressing tool as if you wanted to create a shape half-way between the final form and the flat plate. You then continue pressing with the pressure tool, until finally the formed metal rests against the former. This process does demand a little practice. If the material becomes too brittle while you are working it, it may tear, and you will have to start again, this time annealing the metal again after the second stage. Once the pressing process is completed, you can cut the workpiece to length while it is still on the pressing former. The parts can then be separated. If it they are reluctant to part, moderate heating with a flame will help. Parts 25 and 27 can usually be made without being annealed a second time.

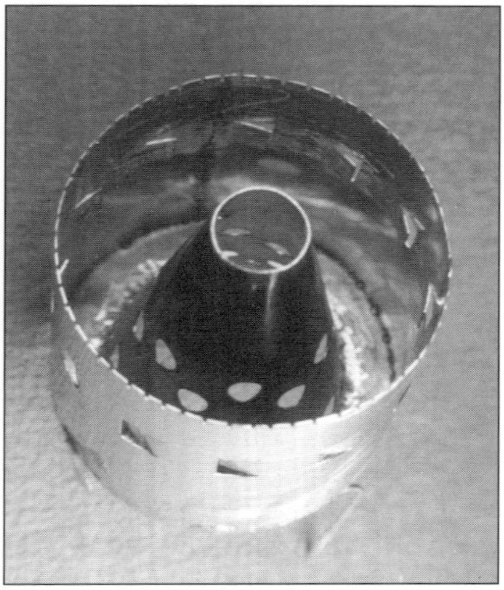

Rear view of the combustion chamber. This component presents no technical or constructional problems. The position and size of the holes and openings are crucial to the quality of combustion. The hole sizes of the version shown here are not exactly as described in the building instructions. The pipe projecting at the bottom is the supplementary gas line.

The same combustion chamber seen from the front.

Once part 25 has been formed, you can glue the reinforcing ring, part 26, to it using UHU Endfest 300. Cure the epoxy with heat. When the glued joint is hard, saw out the central hole to accommodate part 27. It does not matter if this opening is not exactly central. Drill holes in parts 25 and 26 for the retaining screws, part 41, in line with the holes in the internal structure. You can now screw the cover, the internal structure and the housing together, and centre up parts 27, 28 and 25 using the compressor wheel itself. The machined curve in the compressor wheel cover plate serves to locate part 27 accurately. Parts 25, 28 and 27 can now be glued together in a single operation, with part 27 engaging in the opening in the compressor wheel. For once fast-setting epoxy is adequate for the job, although the metal joint surfaces should still be cleaned carefully and keyed with coarse abrasive paper. Take care that no resin gets between part 27 and the compressor wheel. When the resin has cured, separate the parts again (in so far as you have not glued them together). The opening in the compressor wheel can now be turned down on the lathe to produce 0.3 mm clearance between the nozzle and the cover plate both axially and radially. Apply sealing lacquer to the machined opening. It is a good idea to re-check the balance of the compressor wheel after completing this stage.

7.2.9 The combustion chamber

The combustion chamber consists of the inner cone, part 29, the front section 30 and the outer jacket 31. Mark out these parts as shown in the drawing, and drill the holes in part 29. Place a sheet of hardwood under the thin sheet metal before applying the drill. It is important to use the right type of drill with this material. HSS drill bits, in conjunction with stainless steel cutting paste, have proved excellent. Remove rough edges from the holes using a miniature hand-held grinder.

All the sheet metal parts can be cut accurately from the sheet material, without distorting the panels, using a miniature hand-held grinder and fine cutting discs. Part 29 can then be bent to the correct conical shape. One method is to machine a conical wooden former and bend the metal round it. Alternatively, with a little skill it is possible to do the job using a length of dowel about 15 mm in diameter as a former. Clamp the dowel in the vice, and bend the part round it segment by segment, until you achieve the correct conical shape. Weld the seam using the MIG welder and 0.6 mm diameter nickel-chrome wire. Weld part 29 to the front section, part 30, as just described, then weld part 30 to part 31. It is a good idea to make a plywood locating ring to help centre up part 29 relative to part 31 at the rear end. Cut out the air inlet flaps in part 31 using the miniature disc cutter, and bend them inward as shown in the drawing. It is important that the flaps are cut so that they point in the direction of rotation. Cut the cooling air slots as shown in the drawing, using a 1 mm thick grinding disc. The last part of the combustion chamber is the spacers, which are tack-welded in place using the MIG welder, and the supplementary gas inlet tube, part 43, which should be fixed to part 30. The tube can be soldered in place using high melting point silver solder. The inlet tube should only just project into the combustion chamber, otherwise there is a risk that it will melt when the engine is running.

If you wish to install an internal ignition system, fit a threaded glowplug sleeve on the front face, offset by about 60º relative to the inlet pipe, and hard-solder it in place. If you fit the sleeve, remember to check at the final assembly stage that the projecting glowplug head does not foul any of the internal structure components. A cable duct must also be provided for the glowplug cable. The duct can be fitted at any point in the housing in the area between parts 10 and 30. Thick-walled silicone tubing has proved an ideal insulating material for the cable. Make a spring clip from brass sheet for the glowplug contact. Bend the spacer 32 so that it presses lightly against the inner wall when the combustion chamber is fitted into the housing, part 14.

7.2.10 The vaporiser

The vaporiser consists of a 1300 mm length of 5 mm O.D. stainless steel tube with a wall thickness of 0.3 mm. Squash one end of the tube flat and fold the end over in the manner of a toothpaste tube. Seal the end. The next step is to bend it to shape, but it is essential to anneal it beforehand by heating it to red-hot using a gas torch. Let it cool down, then fill it with the finest grade of quartz sand. Tap along the length of the tube with a metallic object to ensure that the sand collects densely in the tube. When the tube is completely full, push a wooden plug into the open end. Now wind the tube into a coil, as shown in the drawing, using a 70 mm O.D. cylinder as a winding former. The direction of winding is important: from the crimped end of the coil, it must wind in the direction of rotation. Bend the first turn of the coil into a ring of 66 mm internal diameter, and position it approximately central relative to the coil. Bend the other end of the coil inward. Remove the wooden plug and tap the sand out of the tube.

Drill five 0.8 mm diameter jet holes in the tube, spaced out at 72°, at right-angles to the plane of the front ring. These jets should then point exactly in the axial direction, towards the compressor end. To improve the mixing effect, bend the holes outward to form cowls, using a piece of 0.8 mm diameter spring steel wire. When using this tool, hold the wire with its end in the plane of the ring, its outside tangent running at an angle of about 20º outwards. With the wire in this position, crimp the tube lightly at a point immediately adjacent to the hole, using a pair of pliers. This results in the vaporised fuel flowing out in a

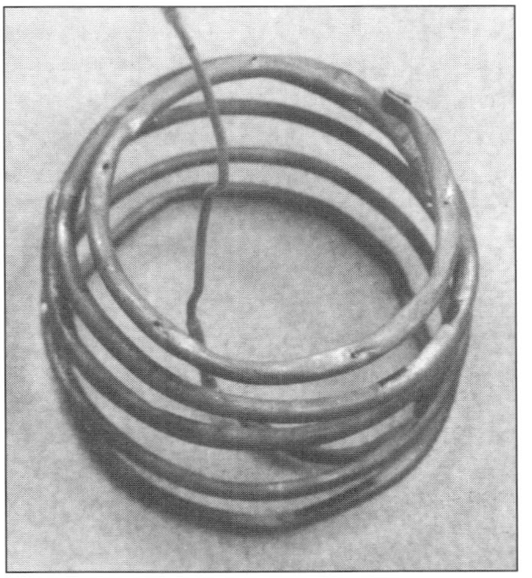

The vaporiser coil, showing the fuel feed pipe and exit jets. The direction of rotation of the coil is the opposite of the version described in the building instructions.

Vaporiser installed in the combustion chamber.

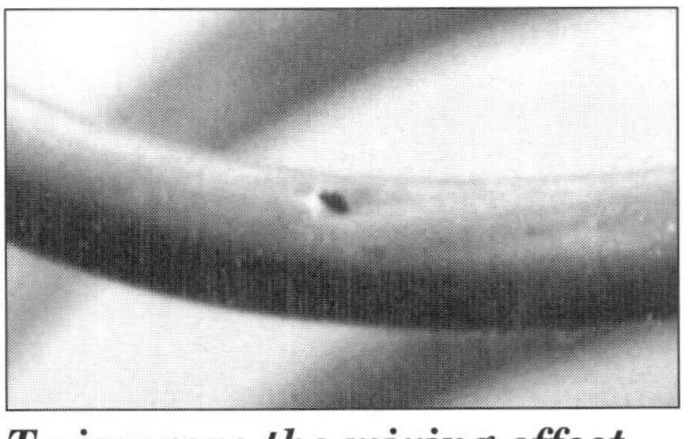

To improve the mixing effect, the holes in the vaporiser should be angled as shown. This is easily done after drilling with a piece of 0.8mm wire.

spiral pattern, which helps to ensure thorough mixing of the fuel and air. Hard-solder the fuel feed tube, part 34, to the free end of the vaporiser tube via an adaptor. Adjust the curvature of the tube so that the solder joint is located in one of the openings in part 29.

If you cannot obtain 5 mm diameter stainless steel tubing of the stated length, you can hard-solder a length of 4 mm stainless steel tube into the main tube to make up the length, although the main tube must be at least 1 m long. The additional solder joint should be located in contact with the outer wall of the combustion chamber. This minimises the danger of the solder joint melting when the engine is running. High melting point hard solder must be used here.

7.2.11 Installing the vaporiser in the combustion chamber

Fit the vaporiser into the combustion chamber from the rear, at the same time slipping the connecting tube forward through the corresponding hole in the inside section. The front ring of the vaporiser is fixed in place with three wire clips made of 0.6 mm diameter nickel-chrome wire (welding rod). Bend the feed pipe so that it runs forward between the blades, between the base plate and the housing wall. Drill a 3 mm diameter hole in the cover at this point. Pass the tube through the hole, and cut it off at a sensible length. Solder a union olive about 3 mm in diameter to the end. Bend the supplementary gas inlet tube to shape in a similar way, and run it to the outside. The combustion chamber and vaporiser are now finished.

7.2.12 Annular jet

This consists of the external jet, part 19, the flow stabiliser, part 20, and the connecting struts, part 21. The fixing straps, parts 22, are also required to fix the jet to the housing. Part 19 consists of a short truncated cone whose outside diameter is exactly the same as the outside diameter of part 18. Part 19 is used to cover this butt joint, and its short cylindrical section should be an accurate fit over the joint. Bend this external part to shape from a strip of sheet metal, tie it tightly round part 18 using wire, and fix it to the tapered part using about 20 spot-welds. Three connecting lugs can now be welded to part 19 and three to part 18, spaced out at 120º. Cut slots in the conical end of part 19 to take the fixing lugs, parts 22, using a rotary cutter. Bend the deflector cone to approximate shape round a length of 12 mm diameter dowel prior to welding the joint. Clean up the weld seam inside and out using the miniature grinder. Make a hardwood cone with a slightly rounded base, and drive the truncated metal cone onto it. If the wooden cone is now clamped in the lathe using a suitable mandrel, you can press the deflector cone to final shape and clean it up. Parts 20, 19 and 21 are assembled as follows: place part 19 on part 18, and wrap nickel-chrome wire round the lugs. With the turbine wheel in position, place part 20 on the turbine disc, using a locating ring and a spacer disc. The struts, part 21, can now be fitted, and parts 19 and 20 joined by means of a few spot welds. Remove the spacer disc and the locating ring, and the annular jet is ready to use. You may find that the fatter end of the deflector cone distorts into a slightly triangular shape, but this has no effect on the annular jet's effectiveness.

The turbo-jet engine has been thoroughly tested with the annular jet held in place by the fixing lugs and wire clips shown, and the system is completely reliable. A gastight joint between jet and housing is not necessary.

7.3 Final assembly

Screw three studs in the M4 threaded holes in the central structure, and fit locknuts on the rear face to secure them. Flatten the studs slightly in the air duct area on the front face, using the disc grinder. Fit the springs which press against the combustion chamber onto the end of the bolts which project on the rear side. You can bend the springs slightly so that they do not slip off the bolts by themselves. There must be at least 1 mm clearance between the projecting end of the studs facing the combustion chamber, and the front face of the combustion chamber itself, to avoid serious stress or distortion occurring when heat causes the combustion chamber to expand.

The front end of the studs is used to secure the cover. Three M4 nuts are sufficient for this. Alternatively you can use two of the studs as a method of mounting the turbo-jet: in this case replace two of the M4 nuts with 7 mm diameter pillars with an internal M4 thread. Fit these pillars through a former at the tail end of the fuselage, and the engine only then requires to be screwed to a suitable support at the rear, by means of the rear front mounting bracket, part 37. If you select this method of mounting, the mounting bracket, part 36, is not required.

Before finally assembling the motor prior to running it, check that all the pipework is unobstructed, and mark each pipe to identify it. Remove any traces of dust and swarf from all the engine's components. Install the front and rear ballraces, and fix the central structure in place using three M3 screws. Clamp the shaft in the lathe chuck and screw the turbine wheel into place by hand, using a cloth to avoid injury. Screw the wheel in place as far as it will go, without using force. Further assembly is completed in the following sequence:

Push part 17 in as far as it will go. Slide the combustion chamber into place and rotate it until the three pipes line up with the holes in the cover. Fit the central structure and line up the threaded holes in parts 11 with the corresponding holes in part 14. Fit the retaining screws but do not tighten them fully. Slide the shaft and turbine wheel into the shaft sleeve from the rear, as far as it will go. The clearance between the turbine wheel and part 18 should now be adjusted by placing three strips of metal 0.3 mm thick between the turbine wheel and part 18, spaced out at 120º. Lightly tighten the retaining screws, part 38. Withdraw the metal strips and check that the turbine wheel rotates freely. I recommend using a feeler gauge, 0.2 to 0.3 mm thick. If the blade can be slipped into place equally easily at all points, then the turbine wheel is correctly centred. Otherwise you will need to make a slight further adjustment, as follows: undo the retaining screw which is closest to the area with the tightest clearance – two of the screws, if necessary – and fit a 0.4 mm feeler gauge at the point with least clearance. Tighten all the screws and withdraw the feeler gauge. If you find it impossible to centre the turbine in this way you will have to make a careful adjustment to the holes in part 14.

Once you have successfully carried out this adjustment, you can fit the compressor wheel and screw it to the shaft using the retaining screw, part 40. Check again that the rotor spins freely. The bearings should be so free that a light puff on the compressor wheel sets it spinning. If the system passes this test satisfactorily, the front section can be fitted and screwed in place. Now repeat the freewheel test again, and re-check the centration of the turbine, as under certain circumstances the housing may change shape when the cover is screwed to it. This may occur if the cover is pressed too tightly against the edge of the housing. If this is the case, grind back the cover slightly where it makes contact with the housing. With the front section screwed in place and the turbine centred perfectly, blow on the rotor with the electric fan, then switch the fan off and hold the engine with the inlet opening facing down. Listen carefully, and you should hear no sounds of rubbing at all. The rotor should slow down gradually – not abruptly. Assuming that any fouling is not due to dirt or excess glue at the edge of the inlet nozzle, the machined opening in the compressor wheel cover plate will need further adjustment.

For the first test runs seal the gap between parts 15 and 14 with a double layer of narrow textile tape, wrapped tightly round over the gap. Temporarily seal the openings between the feed pipes and the cover with thin hose. The engine is now ready for its first run. Don't install the annular jet at this stage.

Carry out test runs of the engine as described in sections 9.3 and 9.4. When you are satisfied that everything works correctly, seal the cover and the feed pipe openings with silicone sealant, using this procedure: remove all traces of oil from the inside of the cover and the edge of the housing. Place the cover on the engine and tighten the retaining nuts lightly, so that the components are in their final position. Now loosen the three retaining nuts by one complete turn, and push the cover forward as far as the nuts allow. Apply a thin line of silicone sealant around the annular gap, and tighten the retaining nuts fully. Seal the pipe openings in the cover with silicone sealant.

7.4 Parts list

Part No.	Description	No. off	Material	Blank Dimensions Notes	Drawing No.
1	Shaft	1		Fabricated item	1
1.1	Central shaft	1	Light alloy	Round bar, 15 Ø	1
1.2	Front bearing spigot	1	Steel C 45 or sim.	Round bar, 12 Ø	1
1.3	Rear bearing spigot	1	Steel C 45 or sim.	Round bar, 12 Ø	1
2	Compressor wheel	1		Fabricated item	2
2.1	Front bush	1	Light alloy	Round bar, 25 Ø	1
2.2	Rear bush	1	Steel C 45 or sim.	Round bar, 25 Ø	1
2.3	Base disc	1	Plywood	6 thick, carbon fibre reinforcement	2, 3
2.4	Cover plate	1	Plywood	6 thick, carbon fibre reinforcement	2, 3
2.5	Compressor wheel blade	11	Plywood	0.8 – 0.9 thick, 3-ply	2
3	Hub	1	Steel C 45 or sim.	Round bar, 12 Ø	4
4	Turbine wheel	1	Cr-Ni steel	2.5 thick	4
5	Front radial ballrace	1	Cr steel	8 I.D. x 16 O.D. x 5, ISO 688	0
6	Rear radial ballrace	1	Cr steel	8 I.D. x 16 O.D. x 5, ISO 688	0
7	Shaft sleeve	1	Steel	Tube 18 x 1 and round bar, 18 Ø	5
8	Flange	1	Steel 37	Sheet, 2 thick	5
9	Strut	3	Steel	Welding rod, 2.5 Ø	5
10	Base plate	1	Light alloy	Sheet, 4 thick	6
11	Link piece	3	Light alloy	Sheet 10 thick, 10x10 or bar 15 Ø	5, 6
12	Compressor blade carrier	1	Plywood	6 thick, fine-ply, beech or birch	6
13	Compressor blade	18	Light alloy	Sheet, 1 thick	5
14	Housing	1	Steel	GAS CV 470 cartridge	7
15	Turbine blade carrier	1	Steel	Sheet, 0.8 thick	8
16	Turbine blade	11	Cr-Ni steel	Sheet, 0.8 – 1 thick	8
17	Central body	1	Steel 37	Round bar 42 Ø or tube + sheet, 0.8	8
18	Turbine housing	1	Steel	Sheet, 0.8 thick	8
19	Annular jet	1	Cr-Ni steel	Sheet, 0.5 thick	9, 12
20	Diffusor cone	1	Cr-Ni steel	Sheet, 0.5 thick	9, 12
21	Strut	3	Cr-Ni steel	Sheet, 0.5 thick	9
22	Loop	6	Cr-Ni steel	Sheet, 0.5 thick	0
23	Stud	3	Steel	Round rod 4 Ø or M4 studding	5
24	Compression spring	3	Spring steel	Wire, 0.6 Ø or similar ready-made	5
25	Cover	1	Aluminium	Sheet, 1 thick	10
26	Reinforcing ring	1	Light alloy	Sheet, 2 thick	10
27	Inlet nozzle	1	Aluminium	Sheet, 0.8 – 1 thick	10
28	Connecting ring	1	Plywood	6 thick	10
29	Comb. chamber, inside	1	Cr-Ni steel	0.5 thick	11
30	Comb. chamber, front	1	Cr-Ni steel	0.5 thick	11
31	Comb. chamber, outside	1	Cr-Ni steel	0.5 thick	12
32	Spacer	3	Cr-Ni steel	0.5 thick	0
33	Vaporiser	1	Cr-Ni steel	Tube, 5 Ø x 0.3, 1300 long	13
34	Fuel feel pipe	1	Brass	Tube, 2 Ø x 0.5 and 3 Ø x 0.5	13
35	Oil pipe	1	Brass	Tube, 2 Ø x 0.5 and 3 Ø x 0.5	0
36	Front mounting bracket	1	Steel	Sheet, 0.8 thick, 2 M4 capt. nuts	7
37	Rear mounting bracket	1	Steel	Sheet, 0.8 thick, 2 M4 capt. nuts	7
38	Screw	3	Steel	M4 x 6 socket-head cap screw	0
39	Reinforcement	3	Steel	Sheet, 0.5 thick	7
40	Retaining screw	1	Steel	M4 x 16 socket-head cap screw	0
41	Screw	3	Steel	M3 x 12 socket-head cap screw	0
42	Supplementary gas pipe	1	Brass	Tube 2 Ø x 0.5 and 3 Ø x 0.5	0

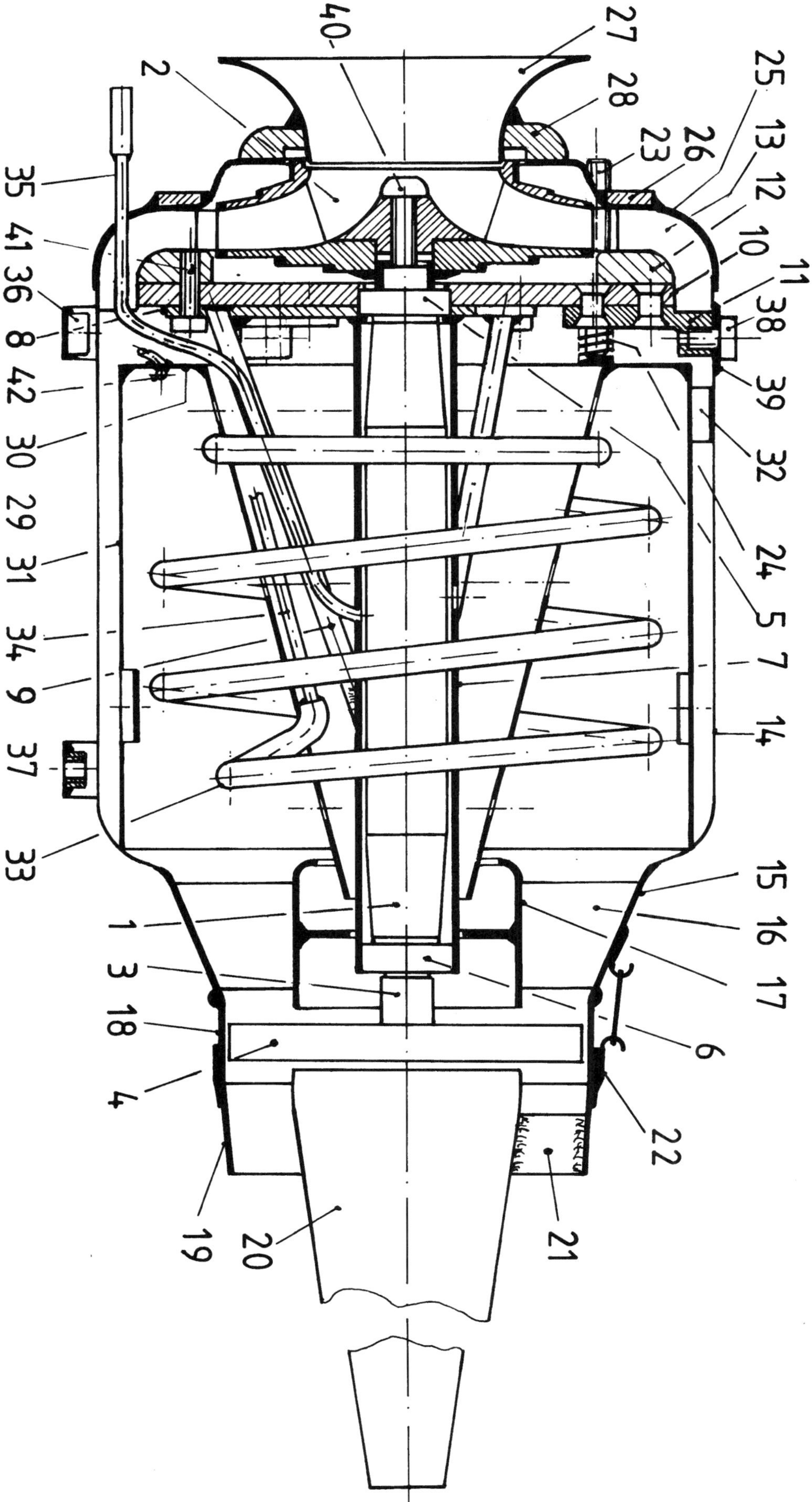

Cross-section through the FD 3/64 turbo-jet

Sheet 1

Part 1,1 Central shaft component

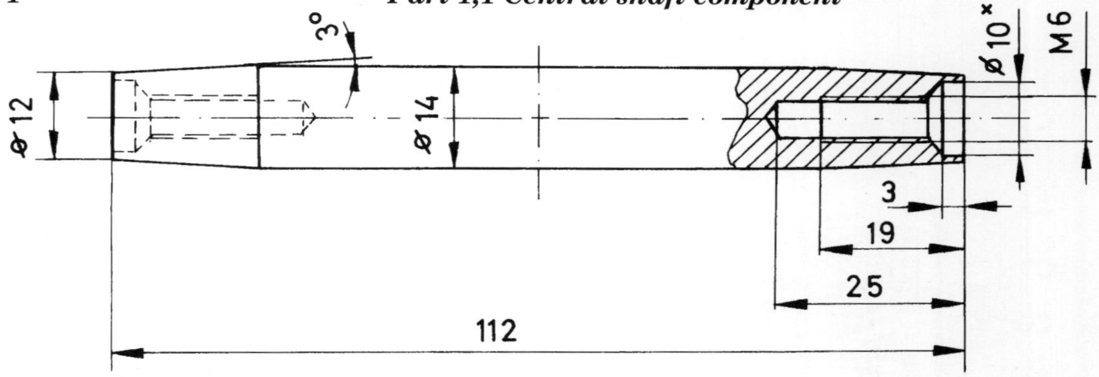

Actual fitted dimensions = see building instructions

Part 1,2

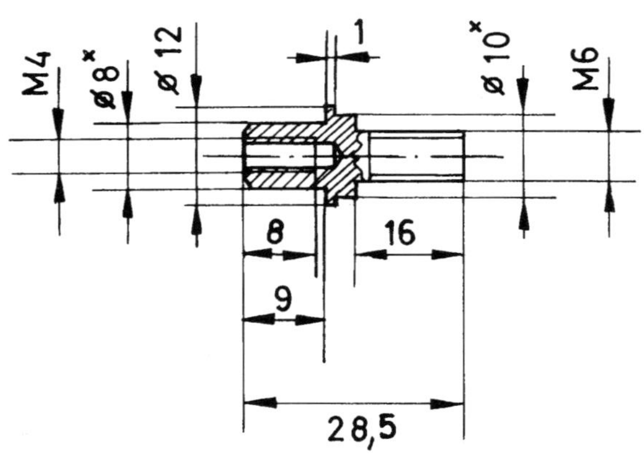

Part 1,3

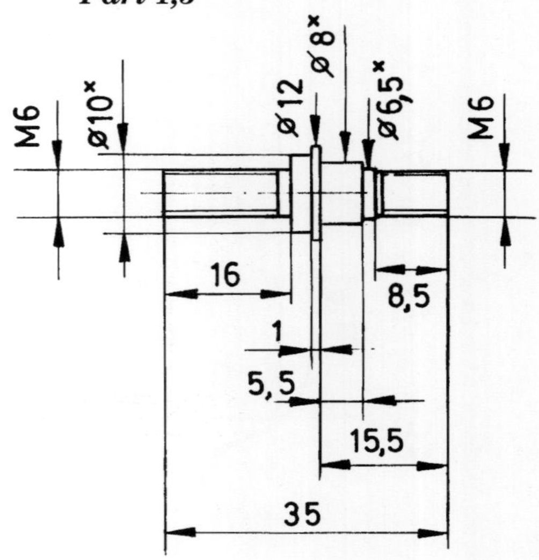

Part 2,1

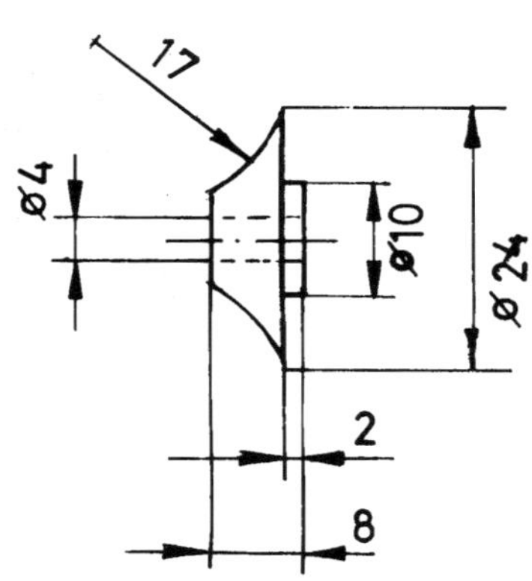

Part 2,2

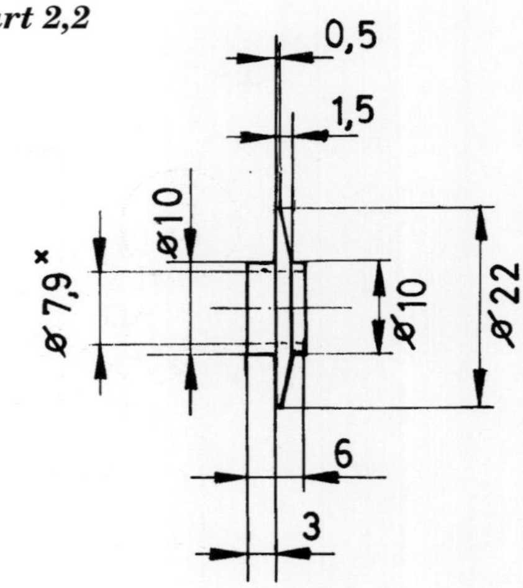

Compressor wheel

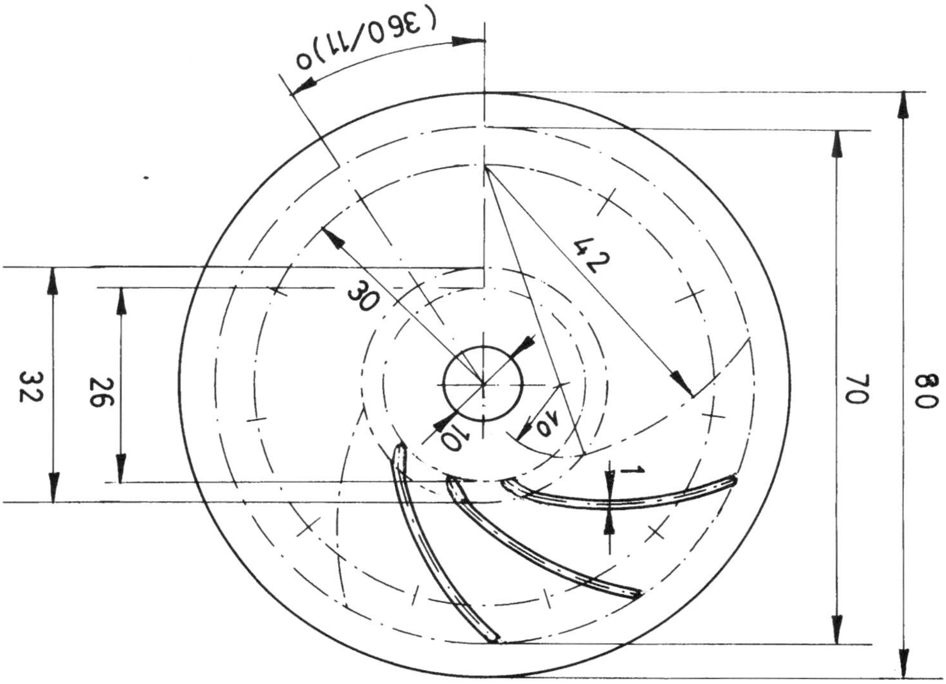

Marking-out diagram for blade slots, seen from the rear of the base disc

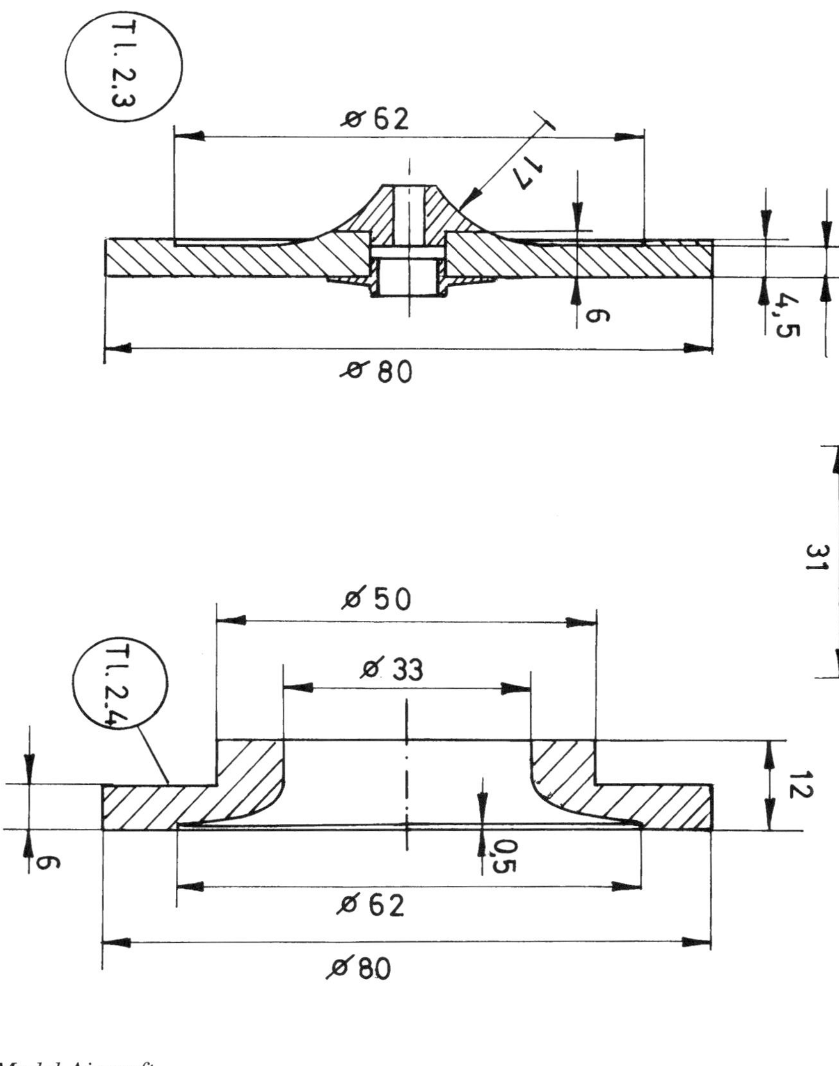

Base disc and cover plate before joining

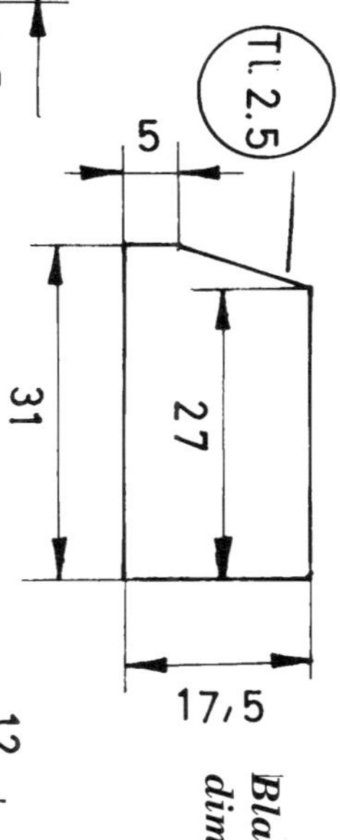

Blade blank dimensions

Sheet 3
Compressor wheel

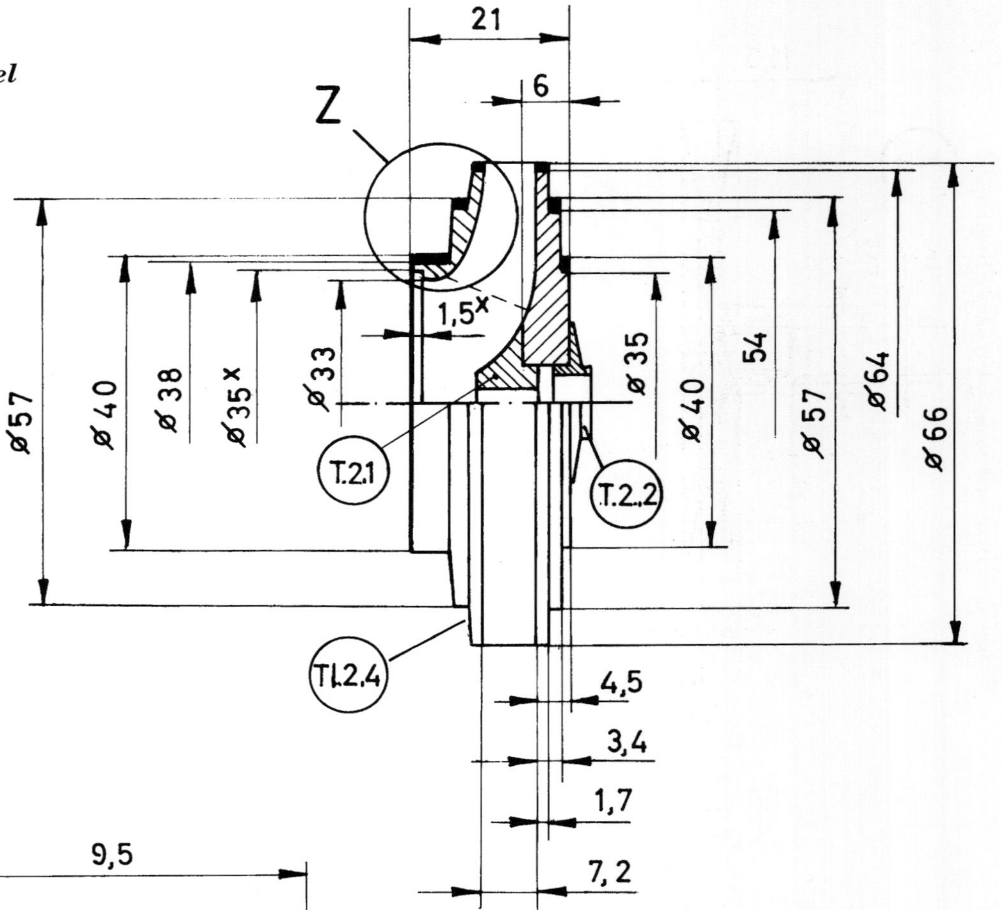

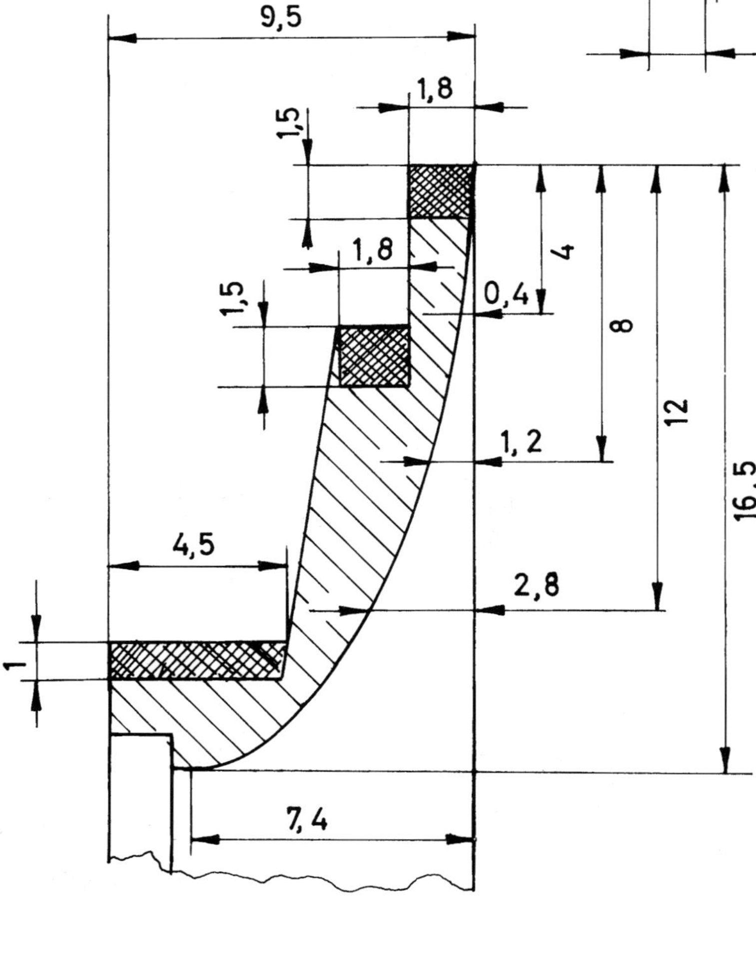

Carbon fibre winding

Detail Z
Scale = 5 : 1

Sheet 4

Turbine wheel

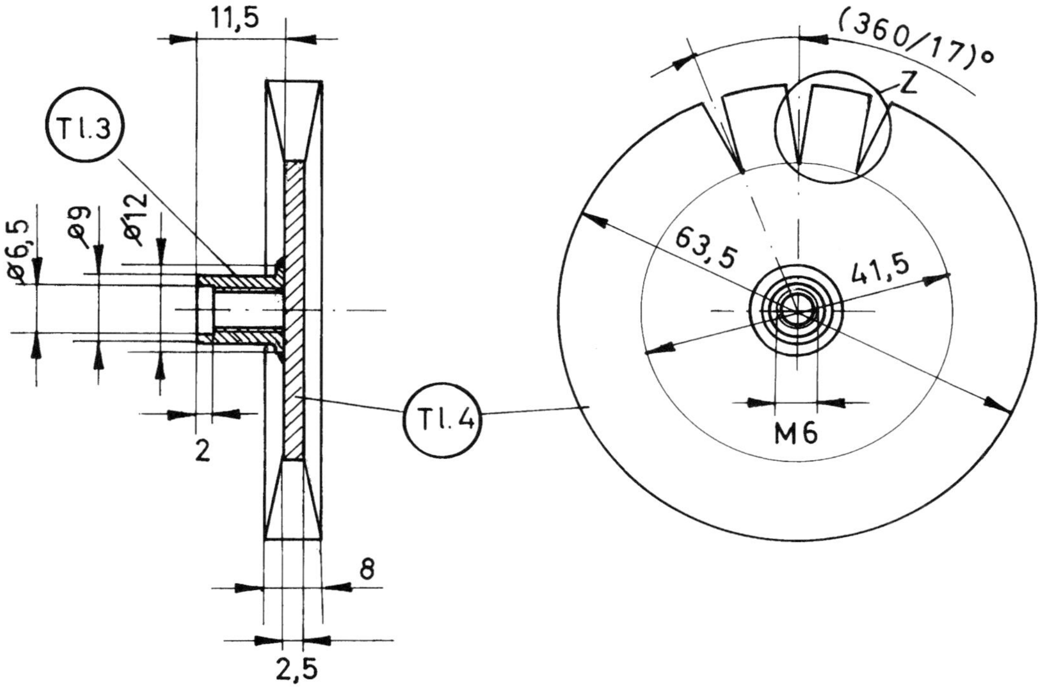

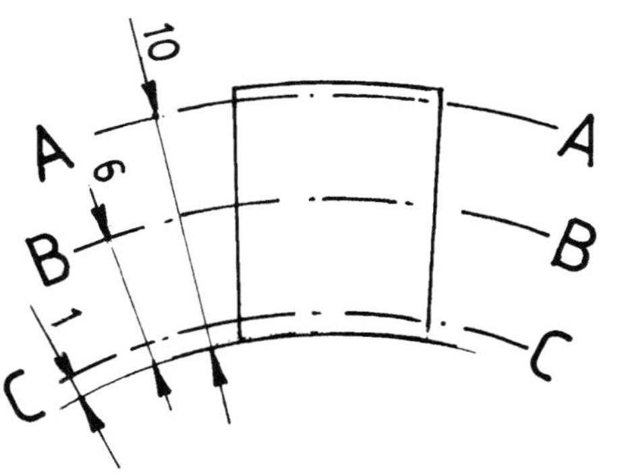

Detail Z
Scale = 2.5 ; 1

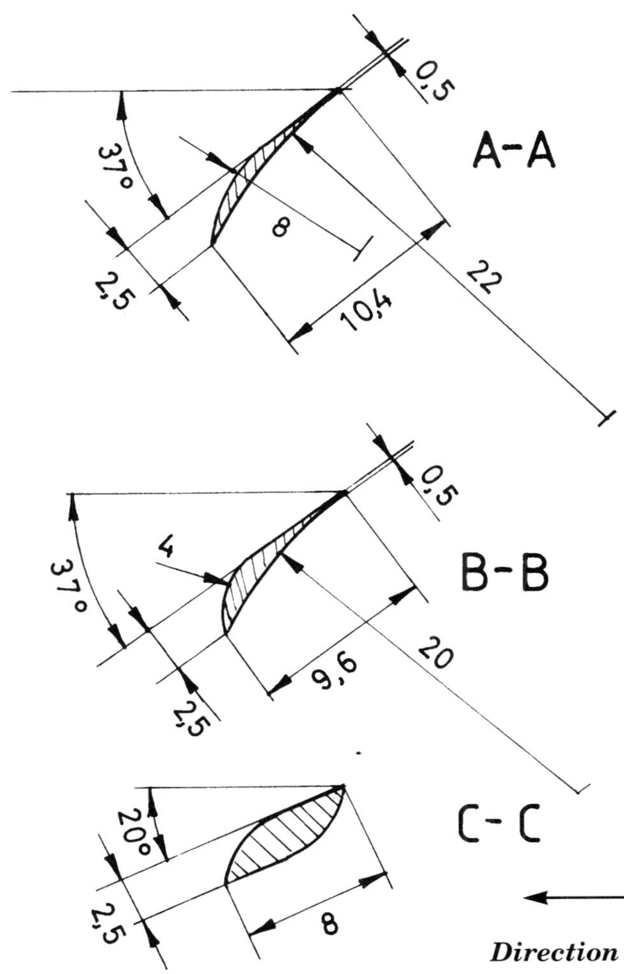

Direction of rotation

Sheet 5

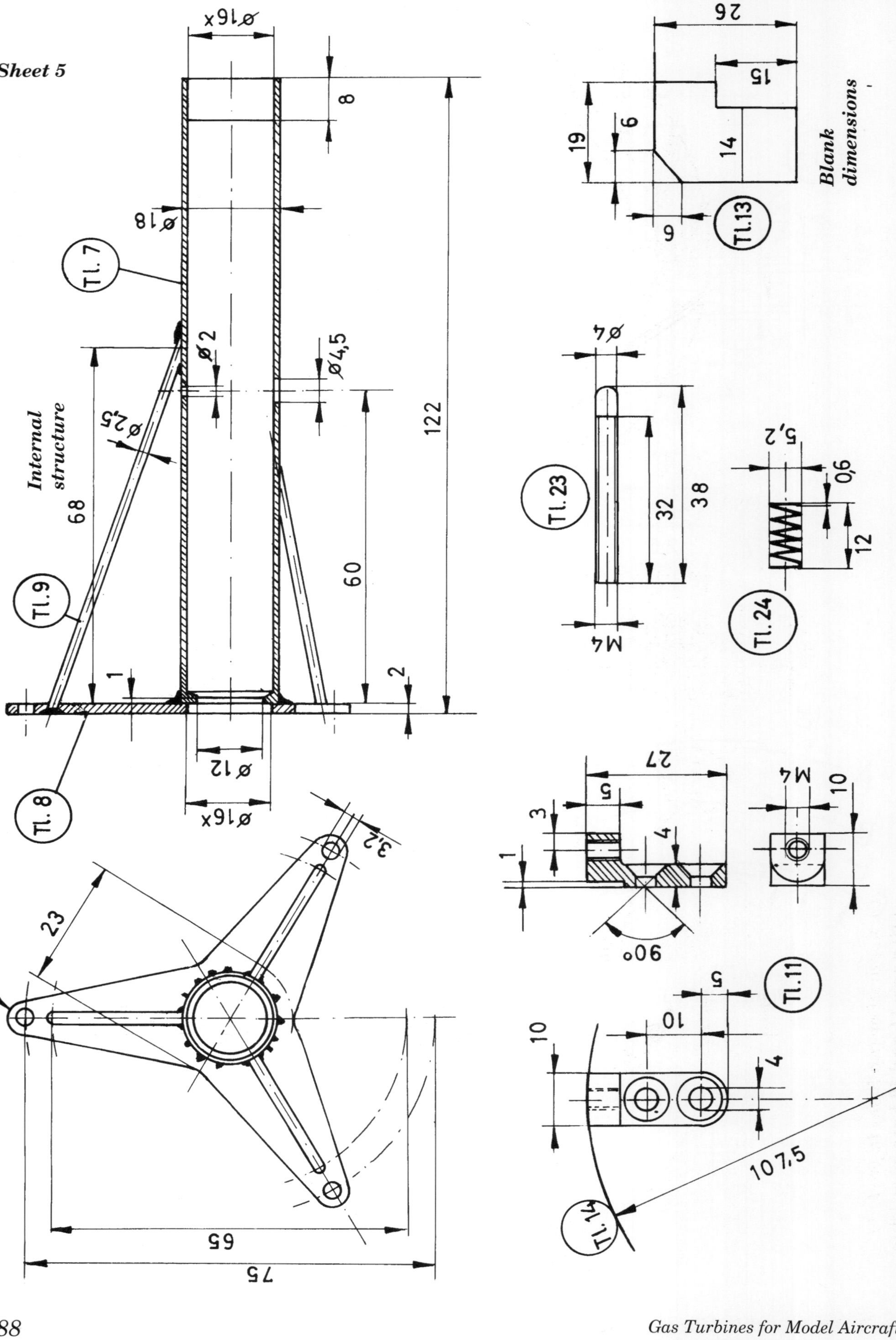

Internal structure
Blank dimensions
Tl.7
Tl.8
Tl.9
Tl.11
Tl.13
Tl.14
Tl.23
Tl.24

Sheet 6

Internal structure

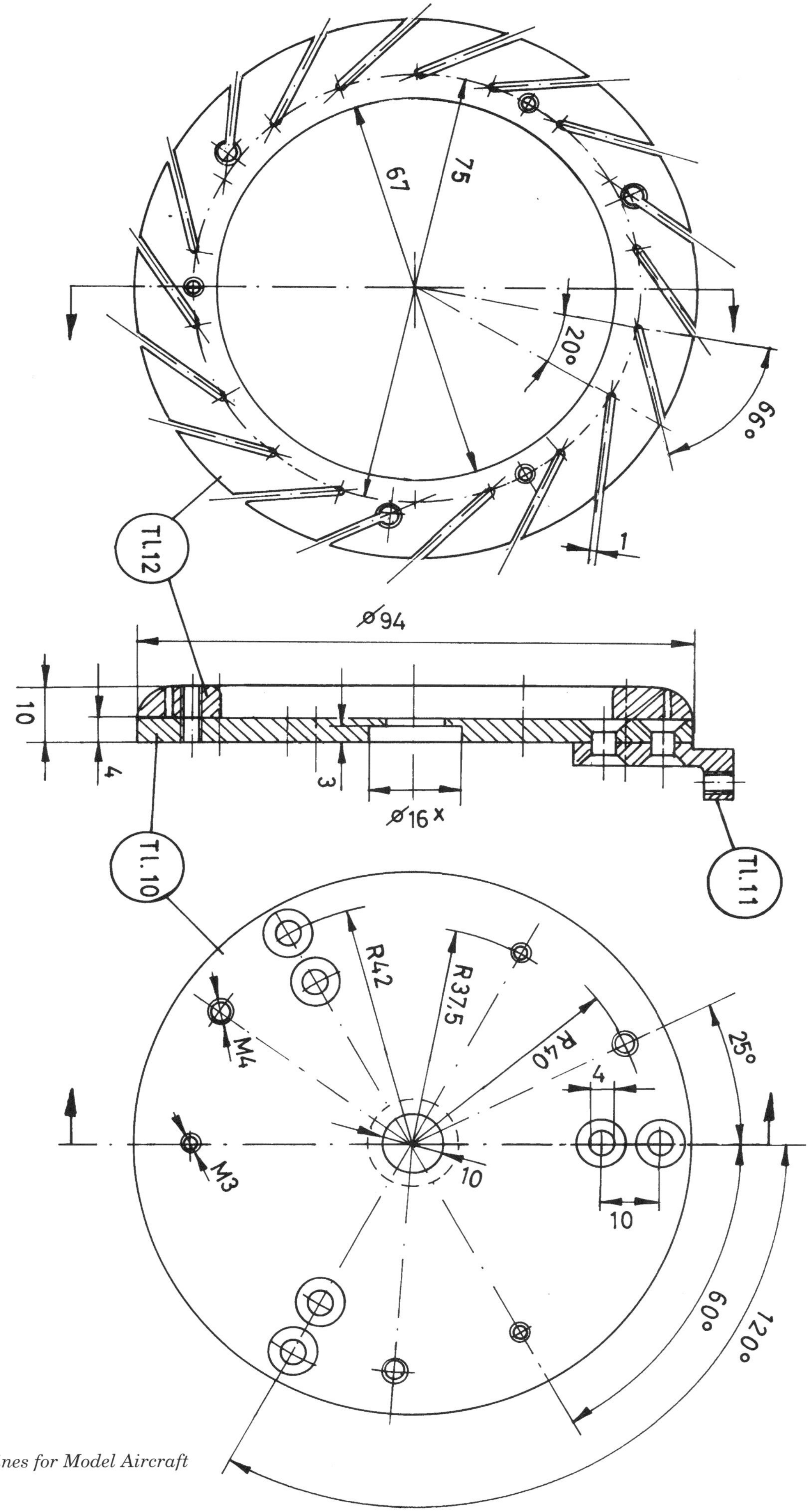

Do not drill or tap holes in parts 11 until parts 12 have been joined.

Sheet 7

Housing

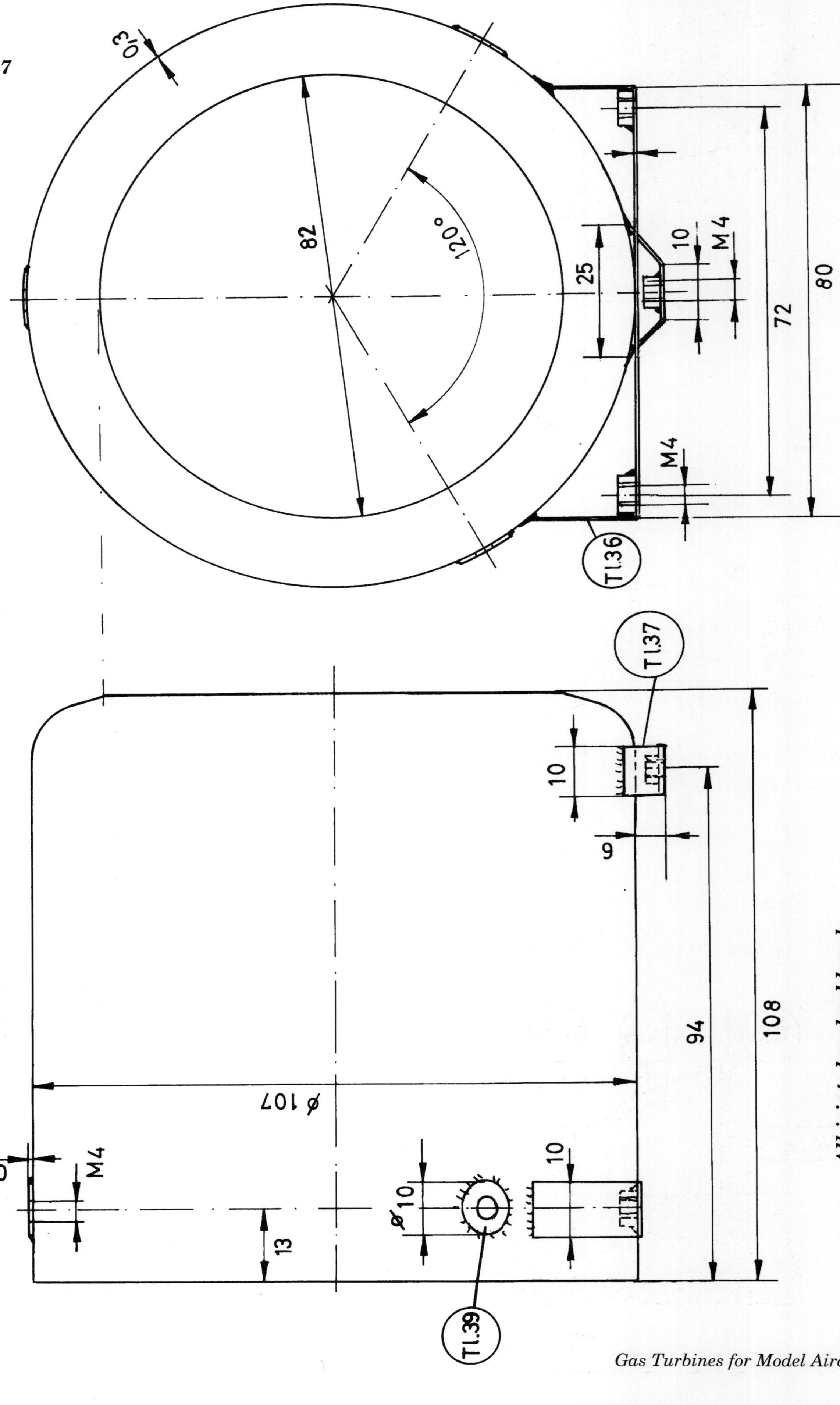

All joints hard-soldered

Sheet 8
Developed view,
parts 15 and 18

Part 15   *Part 18*

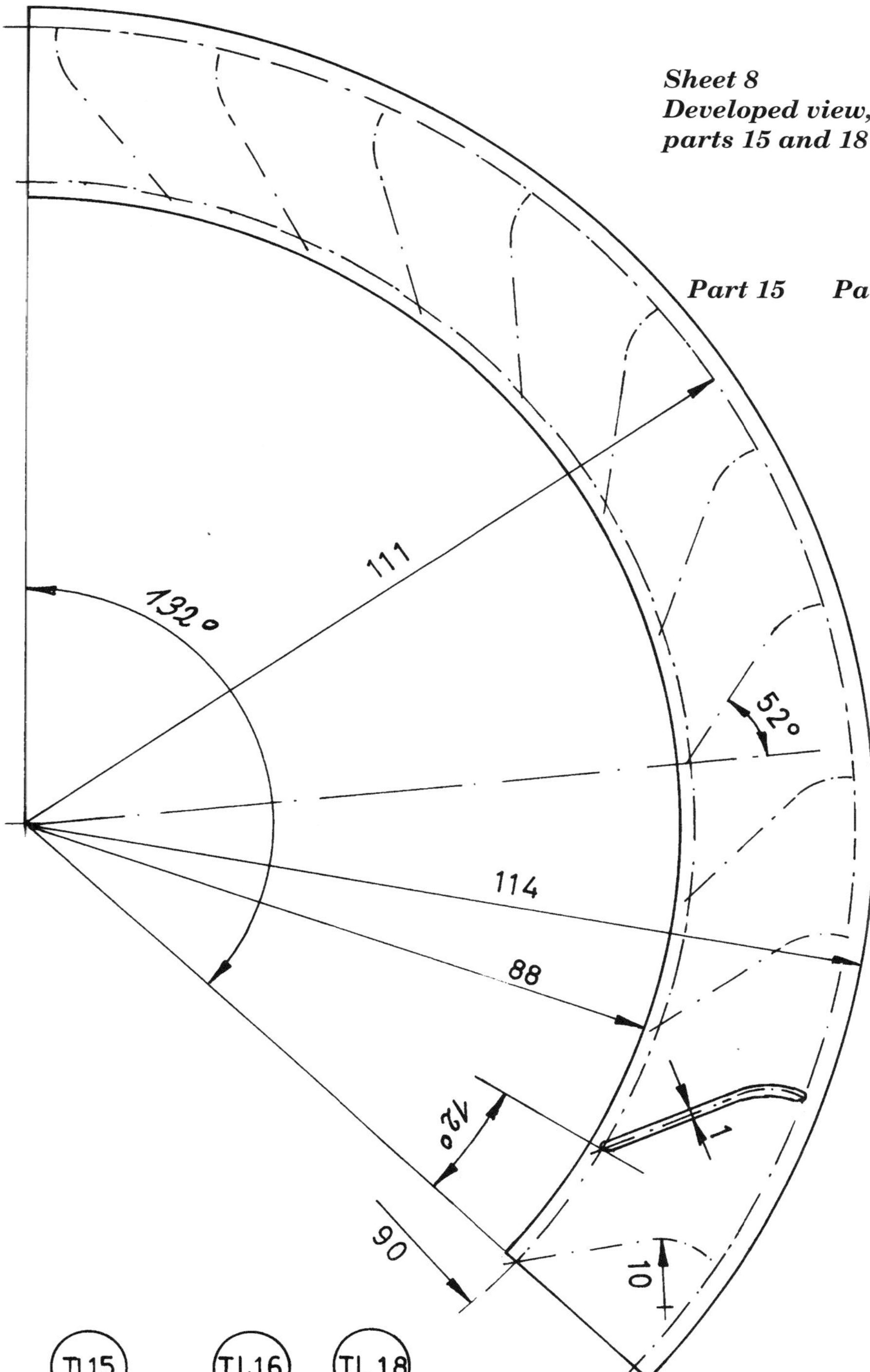

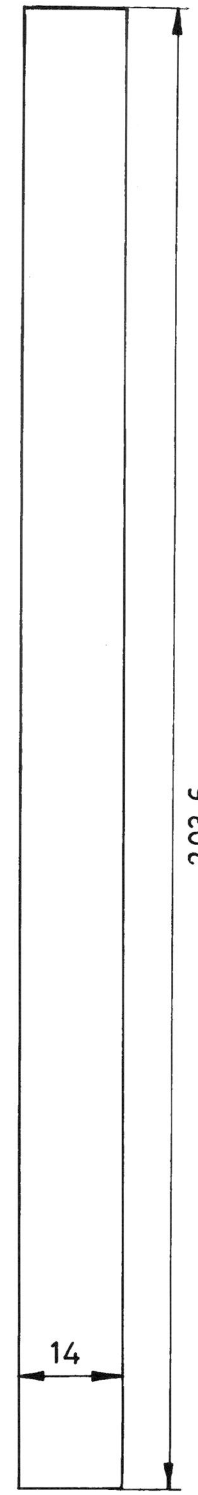

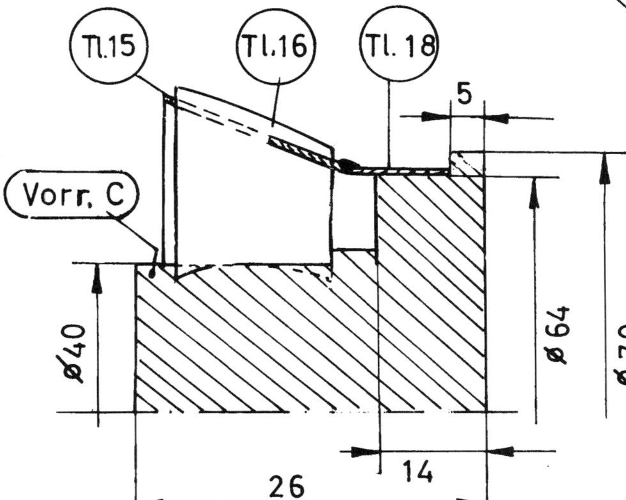

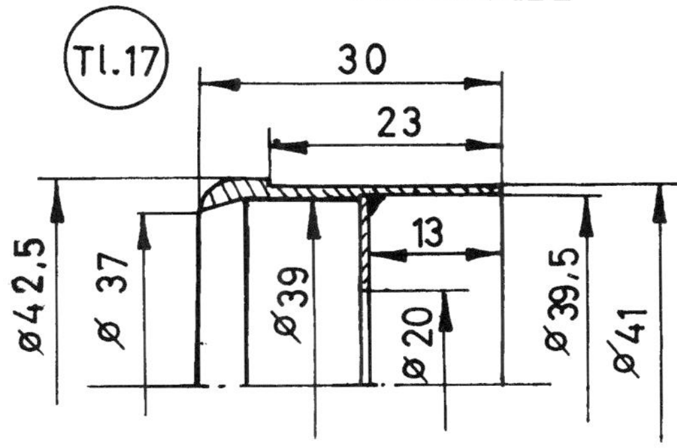

Sheet 9

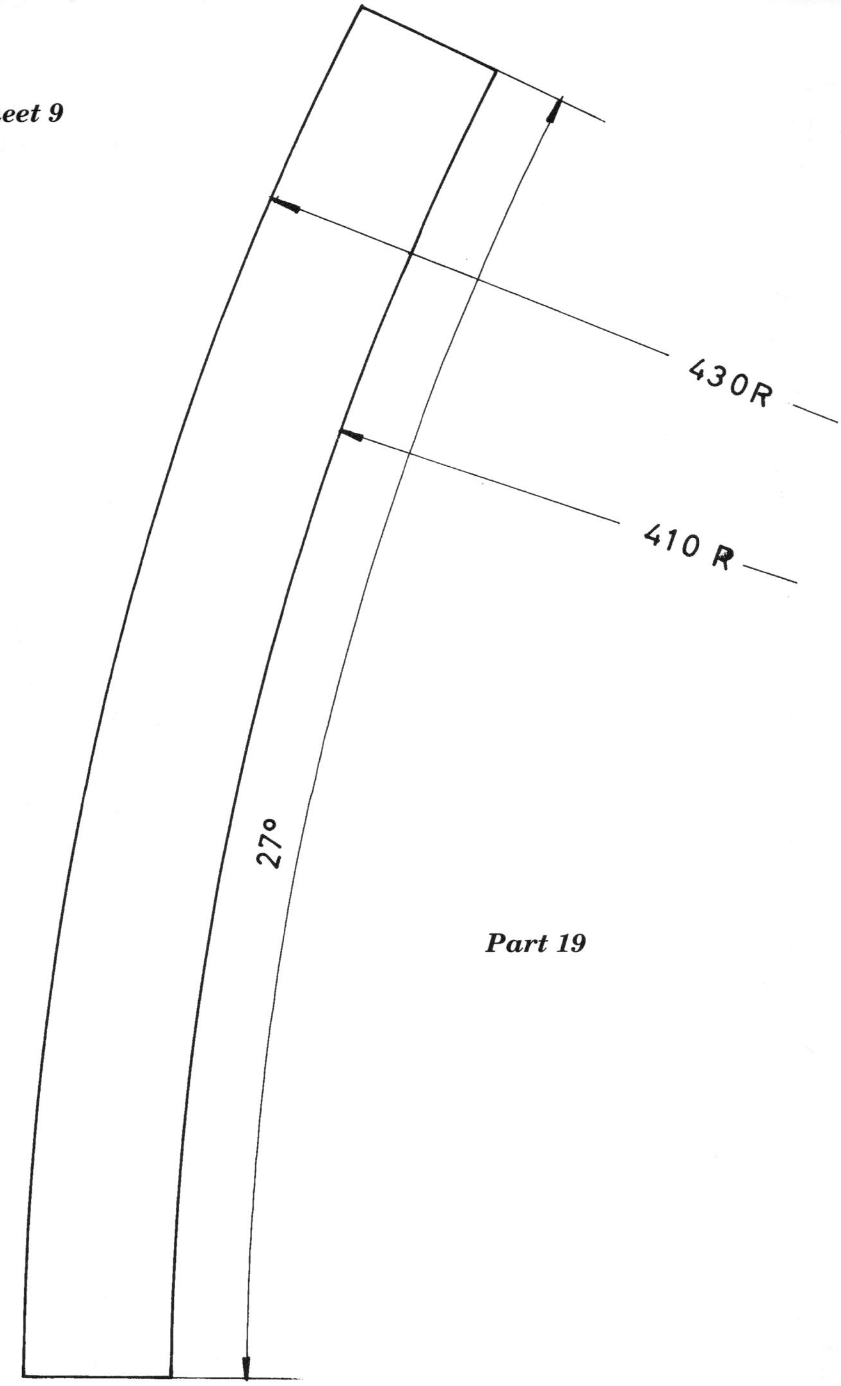

Part 19

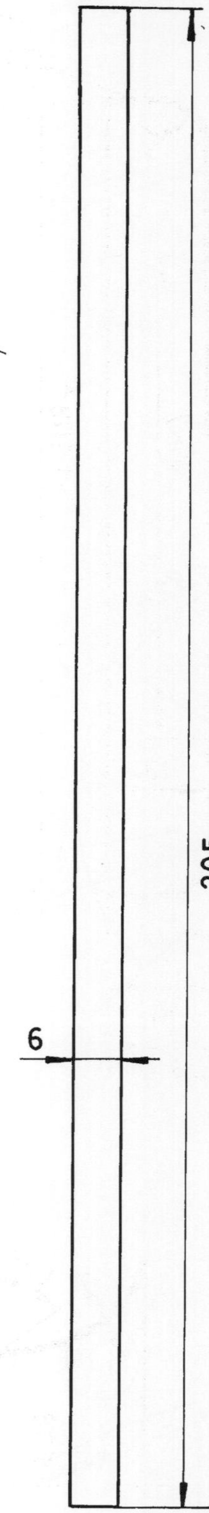

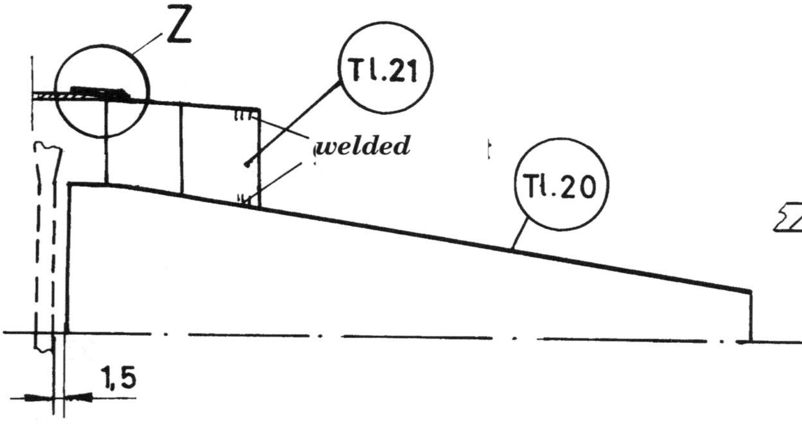

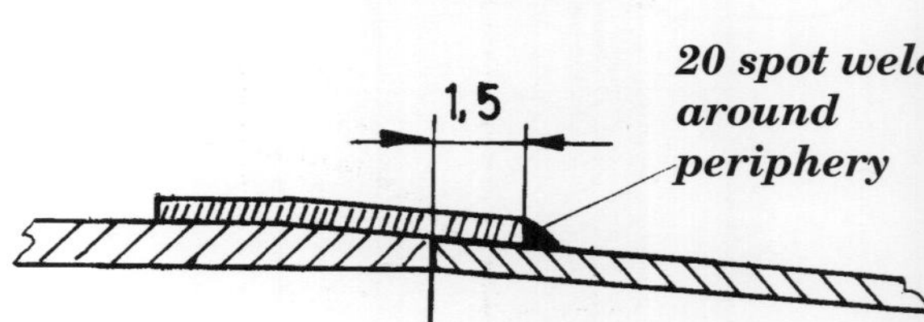

Detail Z
Scale = 5 : 1

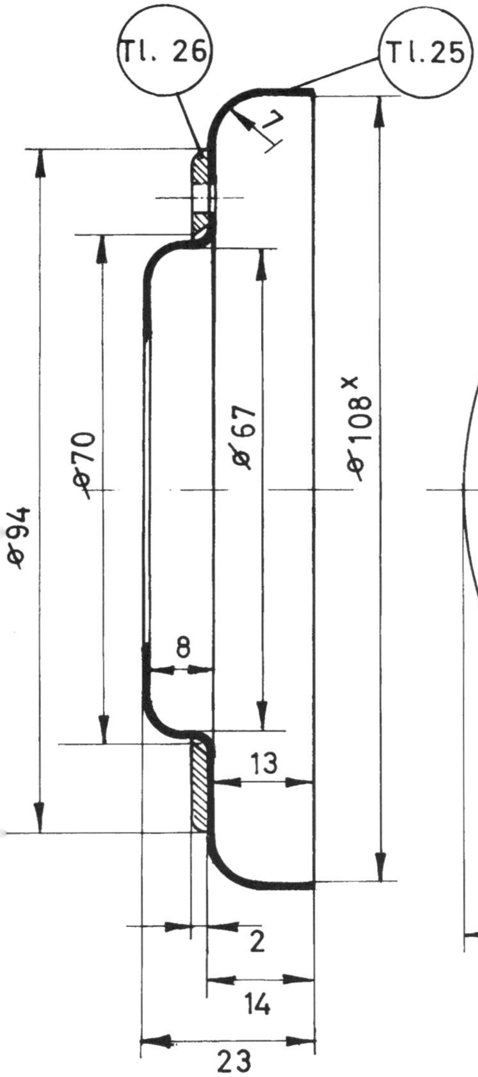
Tl. 26
Tl. 25
1
ø94
ø70
ø67
ø108 x
8
13
2
14
23

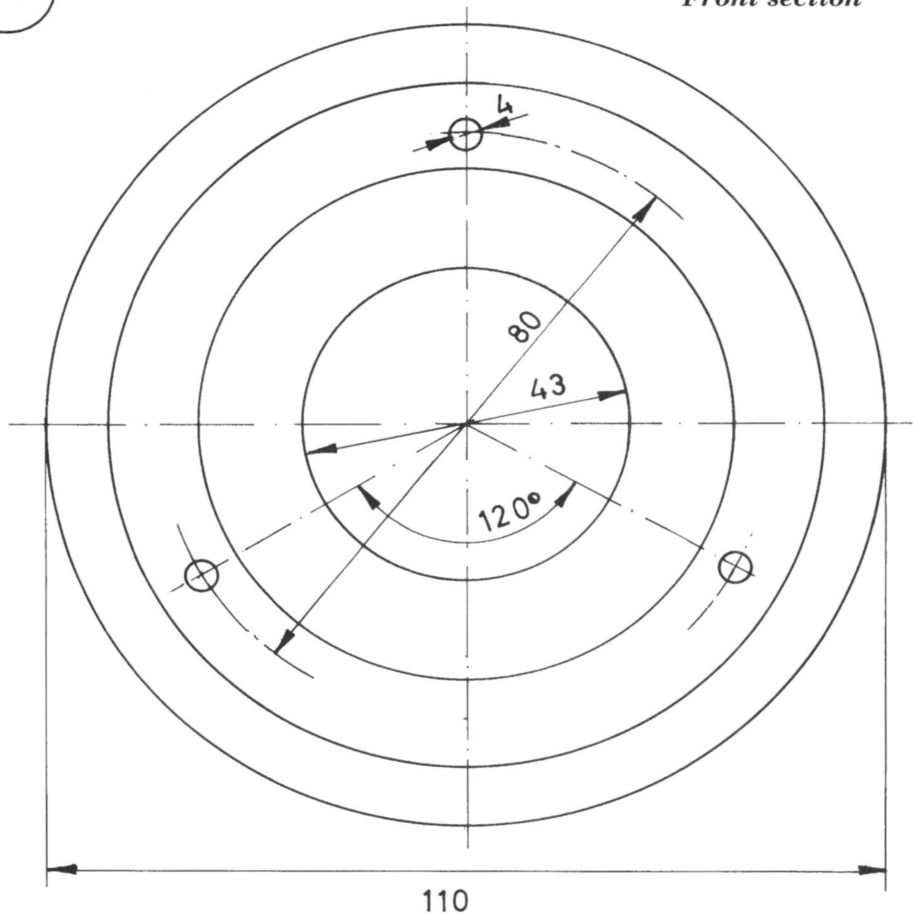
4
80
43
120°
110

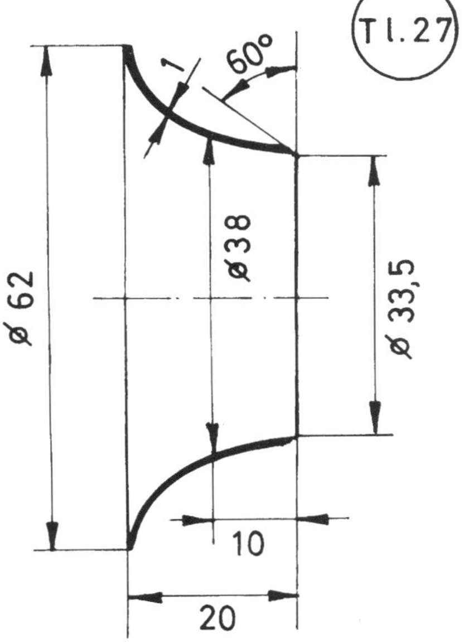
Tl. 27
1
60°
ø 62
ø38
ø 33,5
10
20

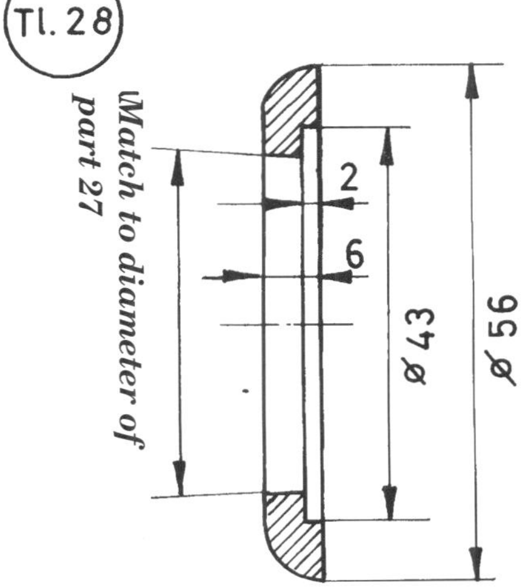
Tl. 28
Match to diameter of part 27
2
6
ø 43
ø 56

Sheet 11
Combustion chamber

Part 29

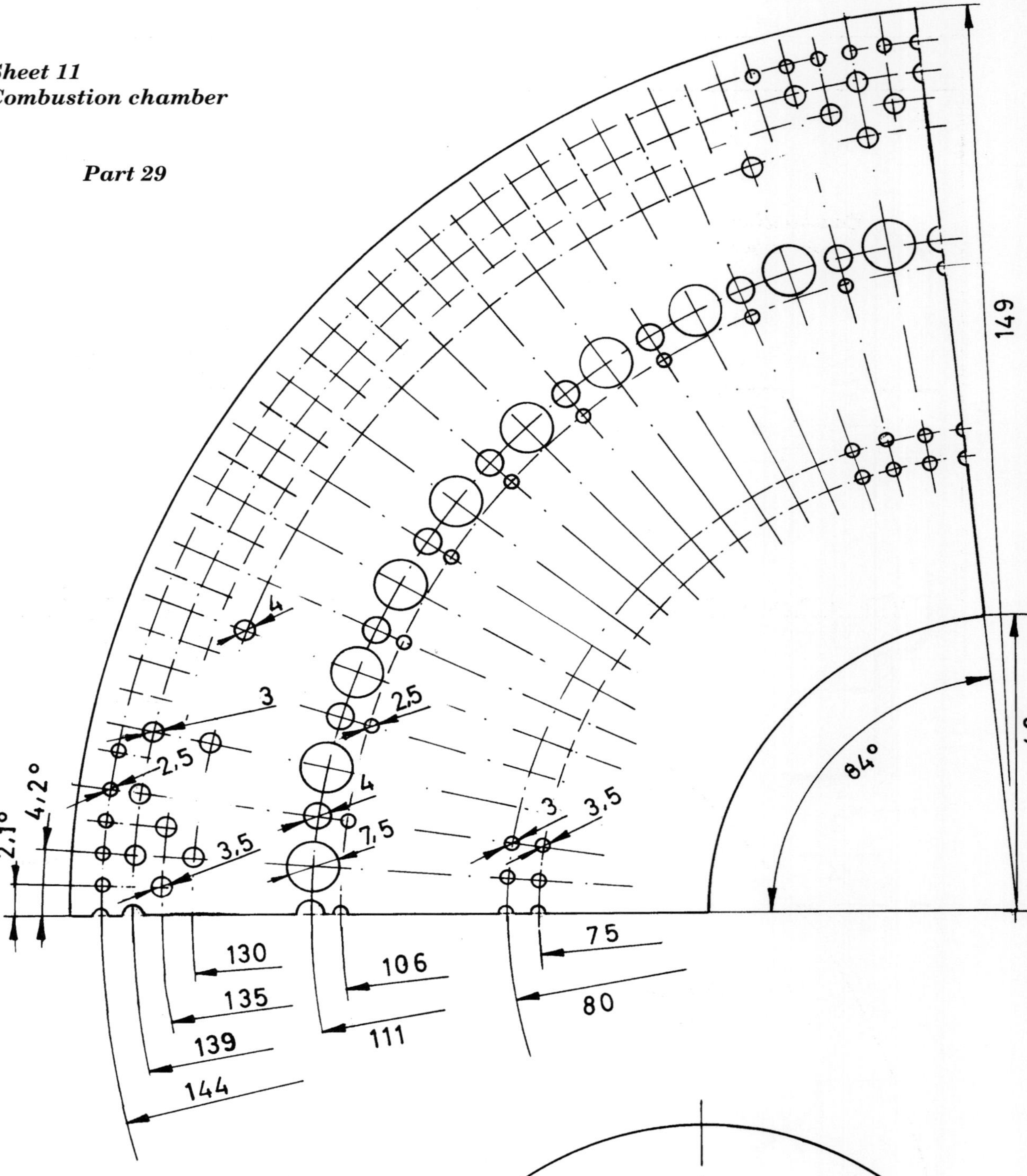

Part 30

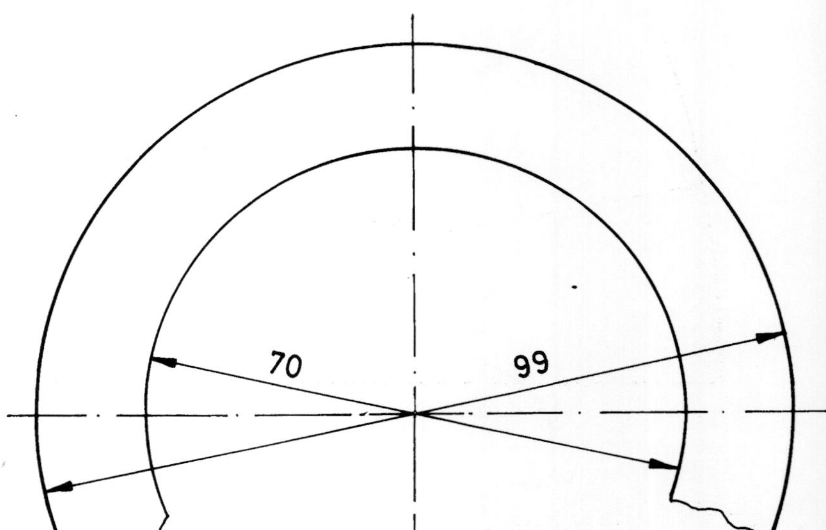

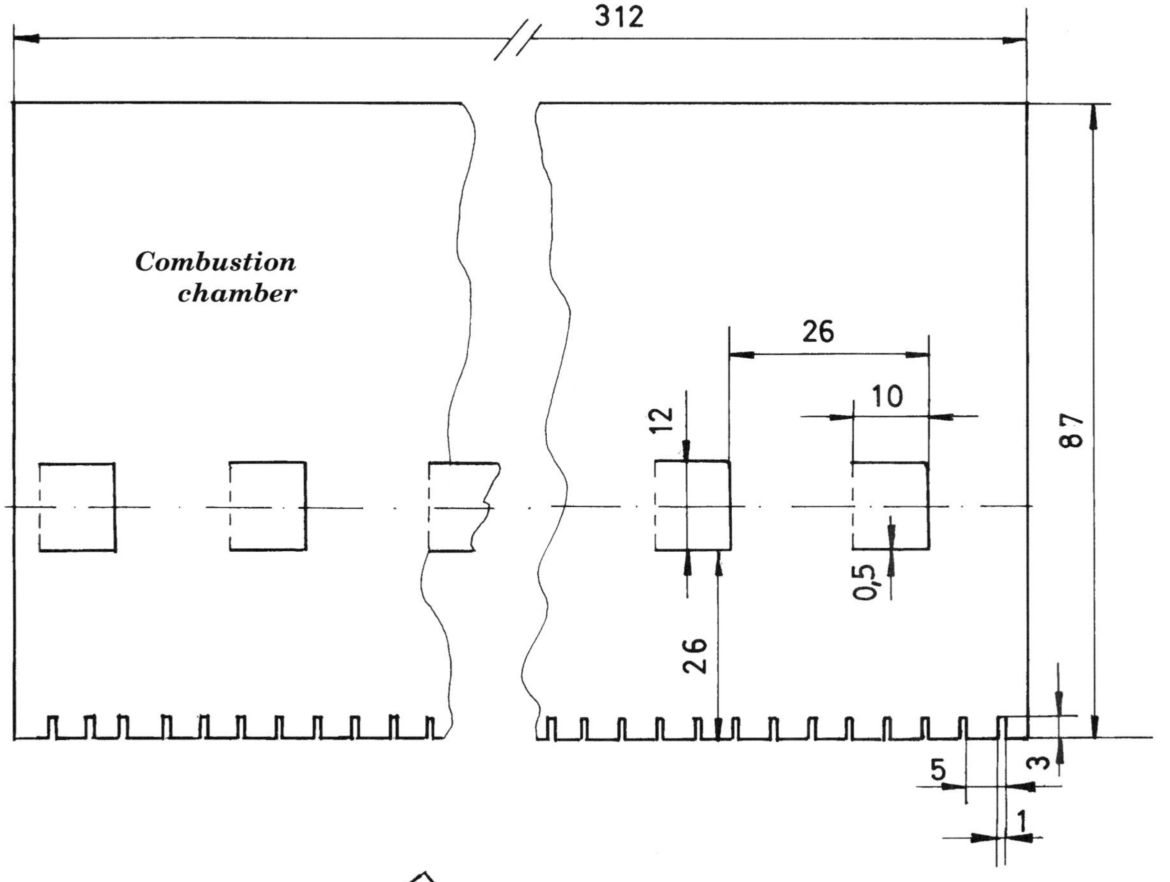
312
Combustion chamber
26
10
12
87
0,5
26
5
3
1

Sheet 12

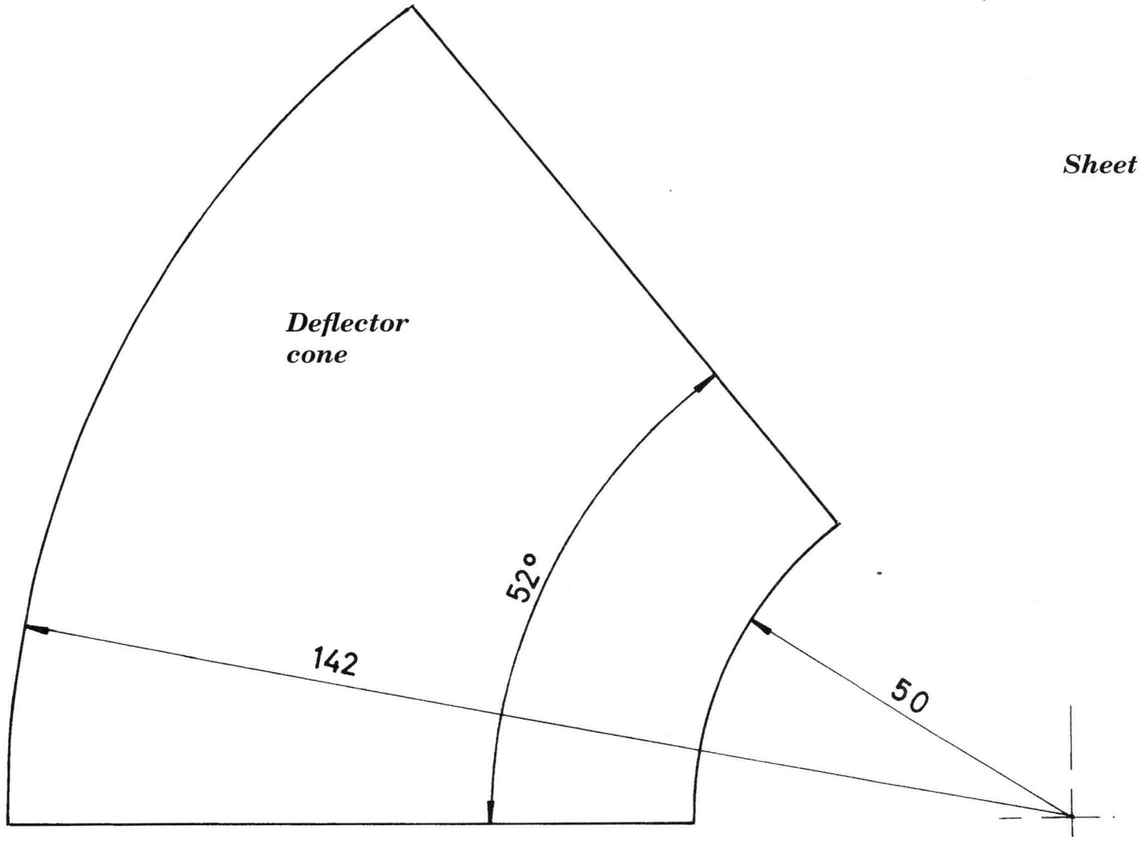
Deflector cone
52°
142
50

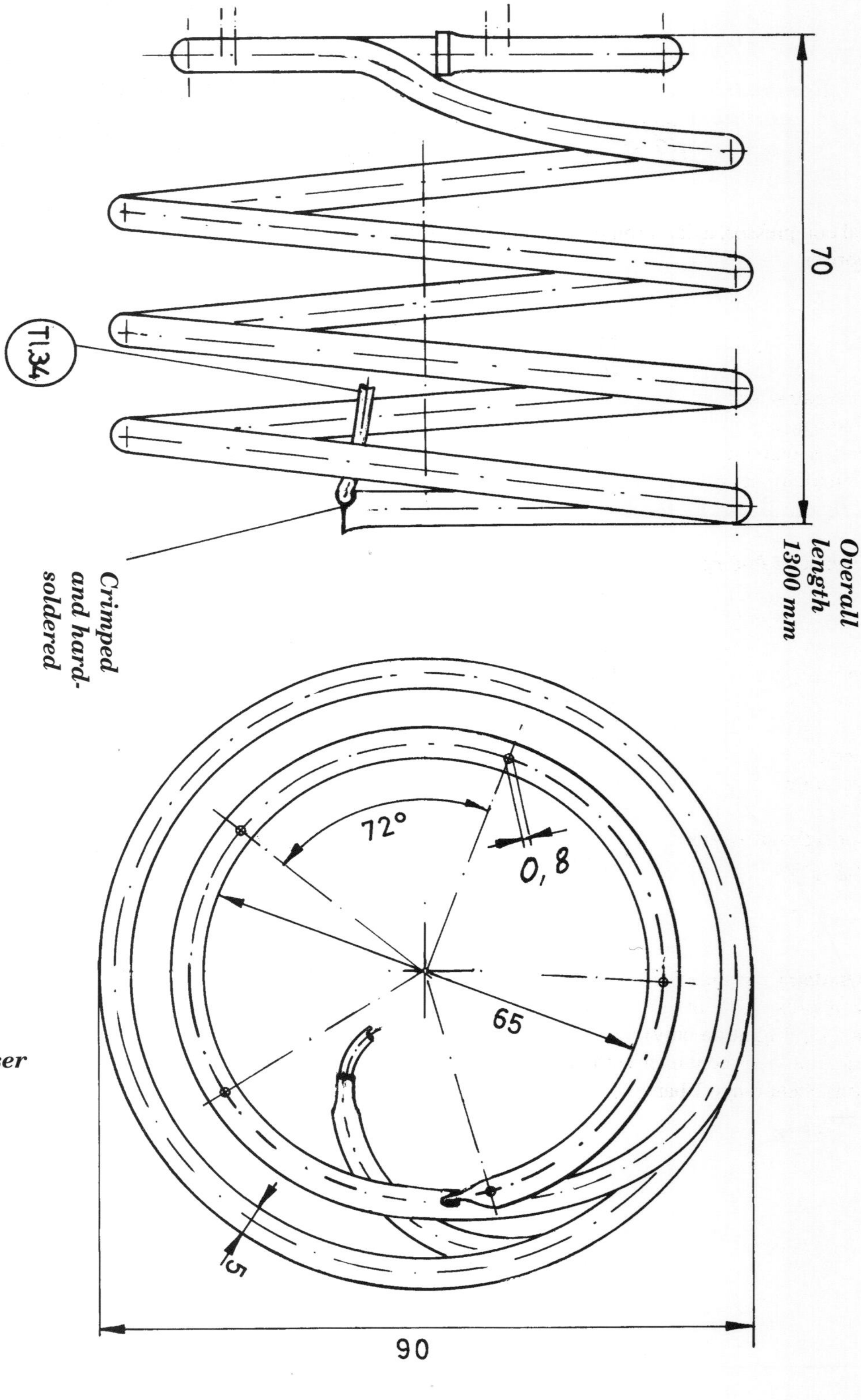

Vaporiser Part 33

Chapter 8

Specification of the FD3/64 Turbo-jet Engine

Basic design:
Single-stage radial compressor, axial turbine, with annular combustion chamber and vaporiser

Dimensions:	
Maximum diameter	110 mm
Length	265 mm
Diameter of compressor wheel	66 mm
Diameter of turbine wheel	63.5 mm
Mass excl. auxiliary equipment	870 g
Mass of airborne auxiliary equipment	280 g

Operational values at 75,000 rpm:	
Thrust	24 N
Pressure ratio	1.4
Air throughput	0.115 kg/s
Fuel consumption	160 ml/min diesel fuel + approx. 10-15% petrol
Oil consumption	2 ml /min
Exhaust efflux speed	209 m/s
Exhaust gas temperature	630º C

Miscellaneous operational values:	
Min-self-sustaining speed	8000 rpm
Idle speed	20,000 rpm
Thrust range	2 - 24 N

Ground-based auxiliary equipment:
Low pressure fan, approx. 20 W motor power
Gas lighter or match (for ignition only)
Propane-Butane auxiliary gas container with outlet valve
Manometer, measurement range 1 bar

Chapter 9

Running Characteristics and Operating Instructions

9.1 The results of an uncontrolled fuel supply

There are two methods of destroying a turbo-jet engine which are so reliable that success is guaranteed. And they don't even include reducing the engine to bits with a lump hammer, or diving into the ground from a great height. These two methods work equally well with any gas turbine and turbo-jet engine, if you handle them incorrectly.

The first is the more spectacular method; variant 1:

Ignite the turbine and run it up until it is just below sustain speed, then switch off the starter and allow the fuel to flow unabated; liquid fuel or gas – it makes no difference. Now, if you try to run a piston engine at too low a speed, it just stops; not so a turbine. The turbine's rotor continues to spin, and rotational speed falls off more or less slowly. At the same time air throughput declines. However, there is still an excess of air in the combustion chamber. If the fuel flow remains constant, the exhaust gas becomes very hot. The turbine blades turn red, yellow, and finally white-hot; the diffusor blades follow suit and, slightly later, the turbine housing and exhaust jet. If we are unlucky, the rotor might seize as a result of thermal distortion, but otherwise the whole lot goes up like a firework display, with brilliant flames and a wonderful display of whizzing sparks. Note please that this only works at its very best if you switch off the starter at the right moment, and leave the throttle setting where it is. It certainly doesn't need full throttle!

And then, just in case variant 1 is not successful, we can turn to variant 2. The basic requirement this time is that the turbine starts without a display of fireworks, and stabilises at its idle speed. A few subdued points of flame during this process won't kill the turbine quickly at this low speed. Now we need a little patience. Even so, the chances of success are quite reasonable, what with experimental development and a lack of experience on the part of the operator. Now we must assume, that for some reason or other the turbine wheel is made to work harder; perhaps by stressed or incorrectly lubricated bearings, or by the turbine blades just fouling the housing. You can hear such effects immediately, and a careful eye on the running temperature and rotational speed will reveal the problem right at the outset. The most reliable method of achieving the wonderful firework display described above is to neglect to switch on the

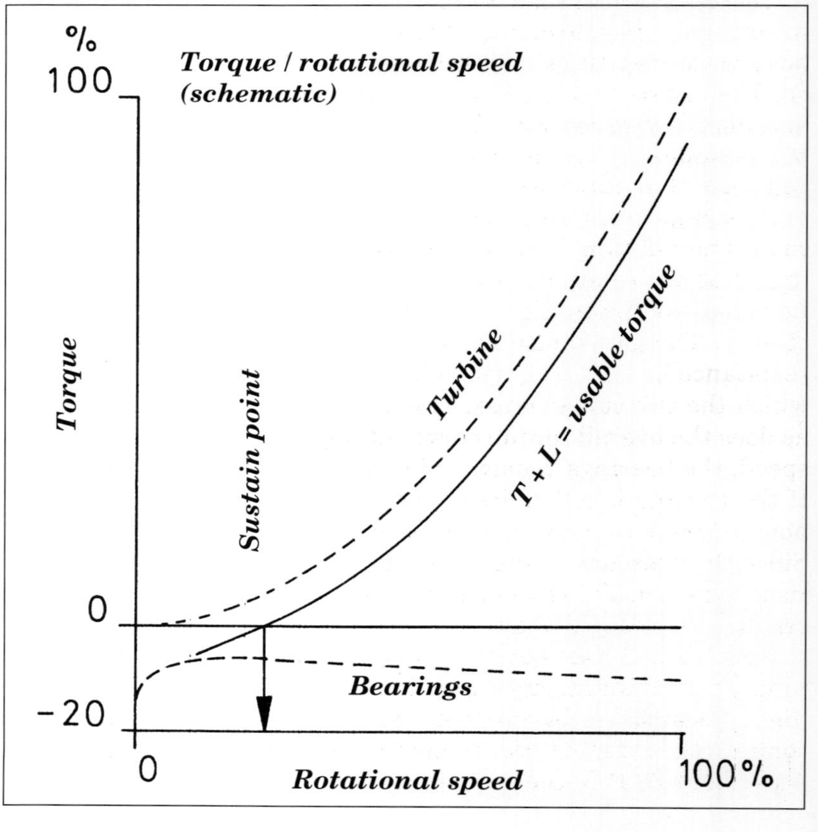

measuring apparatus altogether, and to ignore any unusual sounds which may occur. It is amazing how tenaciously the turbine tries to cling on to life. In most cases the period of time from the first indications of trouble to the sparkling finale is very short, although you can follow it quite well by eye. There are even some experts who follow these exciting events with a video camera. The only drawback to this method is that a new turbine is required for each successful attempt. For reasons of cost I have never felt inclined to let destruction variant 1 run its full course, but variant 2 has proved accidentally successful several times during the engine's three-year development period!

Now we come to the quick method of destruction. The basic requirements just could not be simpler: the most important point to remember is that the fuel pump must be able to deliver fuel in a relentless flow. Otherwise all you do is ignore all the information and recommendations on restricting fuel flow, together with all the other operating instructions.

Dear readers, this is the point at which the fun and laughter stop dead. A gas or jet turbine running at a high and stable operating speed runs away with itself so quickly if the fuel supply is abruptly increased that the first you know of it is the explosion. Every imaginable type and extent of damage and fracture may occur. The energy in the flying shards of a heavy compressor wheel or turbine wheel is comparable to that of shell shrapnel. The preferred direction of flight of these fragments is any direction in the plane of rotation. Their penetrative power varies according to their speed and mass. Cases are known in which the entire rotor has ripped itself out of the turbine housing.

There are also a number of less obvious reasons for a turbine to run away uncontrollably. For example, fuel may collect in the housing during the starting phase, before the mixture is ignited. The excess fuel is spun round as rotational speed increases, and abruptly vaporises. The effect is exactly as if the fuel supply is suddenly increased. Similar things can happen if the vaporiser does not work correctly at low rotational speeds.

It is important to discuss in somewhat greater detail the causes for running away at high rotational speed, and for burning up at low rotational speed when the fuel supply is excessive. Let's look at the turbine wheel's torque curve, which rises very steeply. The graph also shows the torque curve relating to the resistance of the bearings. The sustain speed is the point at which the two curves cross. This point varies with temperature, as does the overall torque curve. At any speed below the sustain speed, the bearings' moment of resistance clearly predominates. If the turbine is left to itself, rotational speed falls off. We can now attempt to convert more energy by opening the throttle. Since the pressure at these low speeds is very low, temperature rises very quickly, but not the working capacity of the gas. The results are outlined above.

Now we will consider the other extreme. The turbine is running at a constant, high rotational speed, so pressure is also high. Because of the large proportion of secondary air in the combustion chamber, the temperature in front of the turbine is only about 600º C. Since there is so much excess oxygen in the exhaust gas, it can suddenly occur that the quantity of fuel being burned doubles or even triples. Naturally this results in a sudden increase in temperature, and at the same time approximately the same degree of increase in torque, because the pressure of the working gas increases. The net result is a dramatic increase in rotational speed. Within a tenth of a second it may rise by 20,000 rpm, i.e. from 75,000 to 95,000. The system is not designed to withstand this speed. Naturally the temperature of the blades rises too, although not as quickly as the increase in rotational speed. If the turbine is already spinning at its maximum permissible speed, the temptation to open the throttle, even for the briefest of periods, should be resisted.

Now we see clearly why it makes sense to set what we term the engine's idle speed considerably higher than its sustain speed. The idle speed should be set to a figure around three times the sustain speed. The ability of the turbine wheel to accelerate is then so good that the engine will run up to maximum speed within a few seconds of the operator opening the throttle. There is a further reason for not running the turbine at too low a speed: we need a certain minimum pressure for the oil supply system, and at the speed stated above this is just adequate.

A typical characteristic of all gas turbines is this: every time you open the throttle the engine's temperature initially rises, then falls back. When the temperature stabilises, we find that the temperature is slightly higher, as shown in the temperature / speed graph. Opening the throttle too smartly is another way of increasing the danger of the turbine blades overheating.

9.2 The influence of air pressure and temperature

In my calculations of thrust, pressure, temperature and fuel consumption I have assumed that the air is at normal temperature and pressure (INA – International Normal Atmosphere). This corresponds to a mean air pressure of 1.013.25 mbar and a temperature of 15º C = 288 K. As we all know, varying climatic conditions involve fluctuations in air pressure and temperature at one and the same location. In an extreme case the variation in pressure may be +/- 5% and in air density +/- 10% either side of the mean value. At a different location we have to allow for the change in altitude. At ground level air pressure falls by 1.2% for each 100 m increase in altitude.

Naturally, these variations influence the operational data of a model turbo-jet engine. Calculating the inter-relationships is a very complex matter, and we don't really need to go to that extreme. However, to be on the safe side I suggest that you observe the following rules of thumb:

a) Influence of temperature

The working temperature of the running turbine rises by about 2.5% for every degree C of temperature rise in the air at the intake. If the air temperature is 30º C, corresponding to a change of +15º C from the "normal" condition, we can expect a temperature rise of almost 40º C at the turbine. If we maintain the fuel supply at the level required for normal conditions, rotational speed rises, but thrust falls off by a few percent. Since the maximum load falls with increasing rotational speed and temperature, we need to reduce the maximum permissible speed when air temperatures are high, by limiting the fuel supply system.

b) Influence of air pressure

With the standard fuel supply rotational speed rises in inverse proportion to air pressure. For example, if you change location from 0m to 500m altitude, air pressure falls by 6%. If all other conditions are unchanged, the engine's rotational speed rises by 6%, i.e. from 75,000 to 79,500 rpm at full load. This means that the engine would be running well above its maximum permissible rotational speed. Thus we must also reduce the maximum fuel supply if air pressure is lower than normal.

To sum up: it is true to say that higher temperature and lower air pressure lead to a lower power output from our turbo-jet, but then this also applies to other internal combustion engines. Changes in air humidity have no significant effect on the operation of a turbo-jet.

9.3 Initial test running: starting and running the engine on propane / butane gas

I strongly recommend these fuels for the initial test run, and also for your own initial experience with running model turbo-jets. I hope it goes without saying that first attempts at running the engine must be made in the open air. The gas bottle must be fitted with an outlet valve which permits fine adjustment. Minimum equipment is as follows: a starter fan or compressed air, a gas lighter, an oil tank, a rev counter, a solid mounting for the turbo-jet, and a mirror.

Fill the oil tank, then force a little oil out of the tank and into the oil line using compressed air (a large syringe). Connect the gas metering valve to the turbo-jet's fuel input using high-pres-

Typical "hot spot". If the exhaust looks like this, combustion is not correct. In this case the cause was a leak in the vaporiser system.

sure hose. Seal the supplementary gas supply pipe with a blind plug. Set up the gas bottle in a location where you can easily operate the valve from a safe position in front of the engine. Place the mirror behind the turbine, about 0.5 m away from it, and angled so that you can observe the turbine clearly from your operating position. The system is now ready to use.

The first step is to run up the engine from the compressor end using the fan, and to measure its rotational speed. With a cold engine this should be at least 3,000 rpm. Switch the fan off, but leave it in the correct position to spin the compressor, ready for use. The rotor's inertia causes it to continue to rotate, its speed declining slowly. Open the gas valve at this point, and apply the gas lighter at the turbine outlet. As speed continues to fall off, the concentration of the gas / air mixture changes, until it inevitably reaches a state in which ignition can occur. If the mixture ignites, but the flame only burns outside the engine, cut back the gas supply slightly. You will know immediately when the flame backfires into the combustion chamber, as the change in the sound of combustion is clearly audible. Immediately switch the fan on again and increase the gas supply, at the same time watching the turbine in the mirror. If everything is working correctly, the engine's rotational speed will rise distinctly. At this point it is very useful if your assistant measures the rotational speed. Once the speed rises above 15,000 rpm, you can switch the fan off. The temperature at the turbine should be no higher than a dark red-hot. If you now the open the gas valve further, rotational speed will continue to rise, and you will detect a slightly higher temperature during the acceleration phase. The liquid gas will tend to vaporise in the supply bottle, and as its temperature falls, the gas flow into the engine also diminishes, and engine speed will fall off slowly. This effect varies very greatly according to the size of the gas bottle.

Because of its low gas pressure, pure butane gas is not suitable for experimental and test work. Propane / butane mixtures may be usable, but their suitability depends very greatly on the ambient temperature. At temperatures below 20º C only pure liquid propane gas is suitable as a test fuel.

It is only fair to expect problems on this first attempt. For example, if you ignite the mixture and you see the turbine glowing brightly, indicating an unexpectedly high rise in temperature, or if you hear any sounds of rubbing or abrasion, shut off the gas supply immediately. It is a good idea to cool the engine down forcibly with the fan. This could also occur when you attempt to speed up the engine. In most cases this problem is caused by the turbine blades fouling the housing. Once the engine has cooled down you will be able to see the traces of abrasion clearly. If they are only on one side, the turbine needs to be re-centred in the housing. If you find several traces of abrasion distributed around the periphery, then the turbine diameter is slightly too large. Turn it down by 0.1 to 0.2 mm and repeat the attempt.

If the vaporiser jets are not evenly adjusted, "hot spots" may show up at the turbine housing or around the diffusor blade area. If this should occur there is no alternative to opening up the engine. This problem can be solved by altering the opening angle of the air flaps in the outside of the combustion chamber, assuming that the problem is not excessive irregularities in the vaporiser jet openings. If the jets are causing the problem, you can try to establish the largest jet opening by carrying out combustion tests using gas and the vaporiser on its own. Try closing the jet partially with nickel-chrome wire to remedy the problem. The way to do this is to push the nickel-chrome wire into the jet, then wind the end round the vaporiser tube.

Once you have overcome these teething troubles, you are ready to increase the gas supply. Set up the gas bottle with the outlet valve on the underside. When you now open the valve, liquid gas flows into the pipe, and thus into the engine. This equates to a substantial increase in the fuel flow, and involves the risk of supplying too much fuel if you are not careful with the valve. When the engine is running on liquid gas you should start measuring the engine's pressure and thrust. Power should only be increased gradually, and it is important to inspect the turbine wheel after each run.

9.4 Starting and running the engine on a diesel / petrol mixture

For this stage you will need the equipment shown in the overall diagram, as described in Chapter 6. The first step is to adjust the restrictor in the fuel line so as to set the maximum permissible quantity of fuel. Start with a wire length of about 10 cm in the restrictor, and measure the fuel throughput with the fuel pump under full load, and the end of the fuel line free. Limit the maximum flow to about 1 ml/s for first attempts.

I recommend a mixture of 70% diesel fuel and 30% standard petrol for initial test running. A small gas cartridge is quite sufficient for the supplementary gas supply. Of course, the gas supply should now be connected to the appropriate pipe on the turbo-jet. The first phase of starting the engine is exactly as described in the preceding section. Ignite the engine and switch on the fan again, then carefully advance the throttle lever on the transmitter. When the fuel reaches the vaporiser, you will see a fairly powerful flame appear at the engine. If rotational speed rises distinctly at the same time, the flame will disappear into the engine after a short period. It is important to measure the rotational speed at this point. If the engine reaches a speed of 20,000 rpm while the fuel flow is kept constant, you can switch off the supplementary gas supply. Engine speed will fall off slightly as you do this. If everything is set up correctly, as described in the preceding chapter, the engine's speed will now rise and fall proportionally when you open and close the throttle. Around the idle speed you will see a yellowish-white combustion through the gaps in the diffusor system, which does not occur when running on gas. Short points of flame at the turbine outlet are not dangerous, but beware of "hot spots".

You can now adjust the idle trim lever to achieve an idle speed of around 20,000 rpm. By carefully moving the trim lever in the direction of full throttle, you can establish the extent to which the fuel restrictor may have to be opened. As already mentioned, you can expect a brief rise in temperature every time you increase engine speed. Before you trim the throttle for the maximum permissible throughput, I recommend that you carry out experiments with an increasingly low proportion of petrol in the fuel. In my experience the ideal petrol content is around 10 – 15%. If you decrease the petrol content further, there is a danger of the vaporiser system oscillating. This manifests itself in the speed of the engine varying periodically over several seconds. If this effect should occur with a high concentration of petrol, then the vaporiser is operating at too low a temperature. The reason for this may be that the coil is positioned too close to the outside edge of the combustion chamber. In this case you will have to remove the vaporiser and bend the coil slightly more tightly. Small kinks or variations in the ideal shape illustrated in these instructions have no effect on the vaporiser's performance.

A further possible cause of oscillation is the use of a fuel pump which is not powerful enough, coupled with a restrictor setting which is too open. Using the pump recommended, any tendency to oscillate can be eliminated by raising the pump's operating voltage and at the same time closing the restrictor.

Initial thrust measurements without the exhaust jet in place will show a maximum thrust about 30 to 40% lower than stated. Once all these tests have been concluded successfully all that remains is to adjust the oil supply rate. This is carried out by altering the length of the capillary tube between oil tank and engine. As a guide you can reckon on an oil consumption of 5 ml/min. at full load.

The final ground test is designed to check the annular exhaust jet. This is primarily a matter of checking that the exhaust temperature does not rise above the maximum permissible value of 630º C. If the temperature rise is excessive with the jet in place, then it will be necessary to open it slightly. The easiest method of doing this is to shorten it by about 5 mm.

9.5 Maintenance

The only parts of the turbo-jet which are subject to substantial wear are the ballraces. Unfortunately it is impossible to state an accurate figure for their useful life. In my experience to date it is safest to replace the ballraces after about every 25 flights, although at that stage no obvious damage is detectable. The same bearings were used in one of the forerunners of the "FD 3/64" whose operating speed was higher than 70,000 rpm. Other points to watch: check for cracks around the blade roots and in the compressor wheel. If you find problems, you will have to replace those parts. One can also imagine that fatigue fractures could occur at the joints in the internal structure, e.g. the welded strut joints. Of course, the useful life of the engine, and especially of the turbine wheel, will depend very greatly on the alternating load due to temperature change. Systematic experiments on this matter have not yet been carried out.

Chapter 10

FD3/67LS, The Turbo-jet from a Kit

The kit for the FD 3/67 LS turbo-jet has been developed to cater for the jet fan who doesn't have the time or the equipment to build a turbo-jet from a plan. The Schneider-Sanchez company, based in St. Lambrecht, Austria, has now been producing the kit since early 1994. As the name implies, this engine represents a development of the FD 3/64, but employing top-quality materials and professional production techniques.

As you will see from the photographs, the kit includes all the turbine components plus the fuel pump and special electronic unit, sensors, valves, filters and fuel pipes. The kit does not include the tanks, the batteries or the starter equipment. The same items as described for the FD 3/64 can be used.

The kit components are supplied almost completely finished and ready to assemble. To produce a functioning turbo-jet all that remains to be done is to remove rough edges from certain parts, solder a number of joints, fit the parts together, fit screws, check centration, and apply silicone sealant seals. The main requirement on the builder is really that he should study the very extensive building and operating instructions carefully. This is not particularly demanding, but it does help if you can read!

Specification of the FD 3/67 system:

Diameter	110 mm
Length	220 mm
Turbine weight	780 g
Mass of auxiliary items	300 g
Max. rotational speed	85,000 rpm
Max. thrust	30 N
Min. thrust	3 N
Acceleration from min. to max. power	2 - 3 seconds
Specific thrust	27 N / Kg
Fuel consumption (max. power) approx.	200 ml / min
Oil consumption	2 ml / min
Fuel	Diesel with 15% petrol or kerosene

Auxiliary equipment:

Fuel pump, electronic unit with temperature and rotational speed sensors

Oil tank, fueltank, 5-cell NC battery (not included in the kit).

The specific thrust of 27 N / Kg, i.e. the ratio of thrust to the weight of the total propulsion system hardware, is unrivalled. As you will see if you read the Chapter entitled "Jet engines and the model", this engine can be used to power some very impressive model aircraft, which are very close in appearance and sound to the full-size prototypes. Several more examples of 'scale' aircraft powered by turbo-jets have been pictured in recent issues of 'Radio Control Jet International', published by Traplet Publications Ltd.

A unique feature of the FD3/67LS model turbo-jet is the monitoring system which keeps a check on the engine's temperature and rpm. The system is based on a specially developed electronic unit, and is designed to prevent the engine running at temperatures and speeds outside its permissible range. The monitoring process begins right at the starting stage.

A rpm sensor in the front face of the turbo-jet constantly sends a speed signal to the control electronics, and simultaneously a temperature sensor sends a temperature signal from the exhaust gas stream. The turbine is started using a fan and propane / butane gas in the usual way. However, the electronic unit does not allow the fuel pump to supply fuel until the engine's speed exceeds the lower limit of about 12,000 rpm, and the exhaust gas temperature reaches at least 300° C. This prevents the danger of liquid fuel being fed into the turbine when it is cold or running at too low a speed. The electronic unit is fitted with differently coloured LEDs which light up to indicate "minimum speed limit reached" and "minimum temperature limit reached" a few seconds after the compressor has been spun up, and the engine has been ignited on the pre-heating supplementary gas. From that point on, the pump's output can be controlled from the transmitter, and engine speed follows the throttle stick position proportionally. The manual valve system provides a straightforward means of switching from supplementary gas to liquid fuel.

The temperature sensor also guards against overheating by switching the pump off instantly if the engine reaches its temperature limit of around 850° C. The engine can then only be re-started if you reset the electronic unit manually. The excess temperature circuit would trip, for example, if you tried to start the turbine on insufficient starter power and excessive supplementary gas. A red warning light glows if the pump is switched off. This also happens if you increase speed too quickly when transferring from supplementary gas to liquid fuel, or if some other problem arises which results in overheating.

The electronic unit also monitors the upper rotational speed limit. If the engine's speed approaches the upper speed limit, the unit reduces the fuel pump's output substantially. This feature compensates for a fuel pump and/or auxiliary battery which deviates from the standard specification. The electronic unit also switches off the pump if the engine's speed falls below the lower limit.

The temperature and rotational speed limits are fixed by the manufacturer. All you need to do is adjust the electronics to match your RC system (zero point, dynamic response). Provided that the engine stays within the permitted operating limits, you can control the electronics, and thus the engine, from the transmitter in the usual way. The engine's construction is straightforward for a turbo-jet, and involves relatively few connections and joints, so that the maintenance is surprisingly easy. The manual supplied covers the procedures in full. However, there can be no doubt that maintenance requires more effort than for the average two-stroke piston engine. In my opinion, and that of the manufacturer, this kit provides the modeller with that degree of technical knowledge which is an absolute minimum requirement for safe operation of a model turbo-jet.

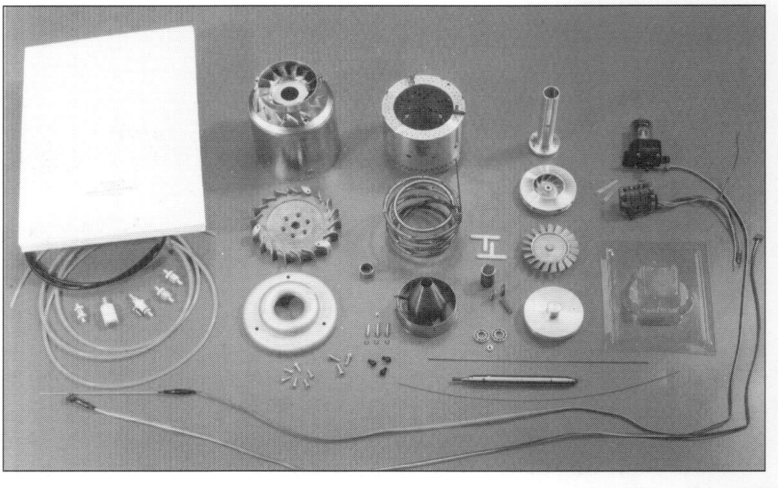

The parts included in the Schneider-Sanchez kit for the FD3/67 LS turbo-jet. (from top left, across and down in order) Operation manual, outer case, combustion chamber, shaft housing, fuel pump, diffuser, vapouriser coil, compressor and turbine wheels, ECU, fuel tube, one-way valves, filter and clunk, front cover, fuel/oil and pressure nipples, rear nozzle, bearings, centring alignment tool, casing for ECU and main shaft. At bottom are the EGT and RPM sensors, attached to their cables.

Please Note: The production version of the FD3/67LS kit is not available from Schneider-Sanchez at the time of printing (November 1994). However, it seems likely that it will become available again in the future, and information on this will be published in 'R/C Jet International' magazine.

Chapter 11
Bibliography

1 Bohl
Ventilatoren (Fans)
Vogel-Buchverlag, Wuerzburg. Germany

2 Dietzel
Turbinen, Pumpen und Verdichter (Turbines, Pumps and Compressors)
Vogel-Buchverlag, Wuerzburg. Germany

3 Kalide
Energieumwandlung in Kraft- und Arbeitsmaschinen (Energy conversion in engines)
An introduction to fluid technology
Hanser-Verlag Munich, Vienna. Austria.

Chapter 12
Sources of Supply

The tools, machines and materials used in the development and construction of the turbo-jet engine as described in these building instructions are, with a few exceptions, available from any good tool shop or engineering supplies company, or by mail order. The exceptions are sheet nickel-chrome steel for the turbine wheel. This material, also known as V2A, V4A and REMANIT, can be procured in the form of scrap material from fitters' workshops and manufacturers of containers such as ovens and heating systems. Metal working companies are generally prepared to supply scrap pieces of round steel and aluminium alloy bar. The "Yellow Pages" will yield the address of your nearest supplier.

Tools and machines, and also thin sheet steel, stainless steel and aluminium are available through DIY shops and engineering suppliers. Any camping supplier will sell you propane gas bottles of various sizes and gas cartridges with matching accessories.

A useful address for measuring devices, as well as all current modelling supplies, is: Conrad-Electronic, Klaus-Conrad-Str. 1, 92240 Hirschau, Germany.

The vaporiser is made of thin-walled stainless steel tubing, which is available from model shops specialising in model boat accessories. This tubing is the preferred material for railings and other fittings on model boats.

A kit for the FD 3/67 LS turbo-jet engine, and a range of matching accessories, are available from Schneider-Sanchez GmbH, Am Gruenen Weg 5, A-8813 St. Lambrecht, Austria. *(See note page 104)*